AN EQUATION THAT CHANGED THE WORLD

NEWTON, EINSTEIN,
AND THE THEORY OF RELATIVITY

HARALD FRITZSCH

TRANSLATED BY
KARIN HEUSCH

THE
UNIVERSITY
OF CHICAGO
PRESS
*CHICAGO AND
LONDON*

The University of Chicago Press, Chicago 60637
The University of Chicago Press, Ltd., London

Paperback edition 1997
Printed in the United States of America
03 02 01 00 99 98 97 2 3 4 5
ISBN: 0-226-26557-9 (cloth)
ISBN: 0-226-26558-7 (paperback)

Originally published as *Eine Formel verändert die Welt: Newton, Einstein und die Relativitätstheorie*, © R. Piper GmbH & Co. KG, München 1988. The translation from the German edition was subsidized by Inter Nationes, Bonn.

Library of Congress Cataloging-in-Publication Data

Fritzsch, Harald, 1943–
[Formel verändert die Welt. English]
An equation that changed the world : Newton, Einstein, and the theory of relativity / Harald Fritzsch : translated by Karin Heusch.
p. cm.
Translation of: Eine Formel verändert die Welt.
Includes bibliographical references and index.
1. Relativity (Physics)—Popular works. 2. Physicists—Interviews. I. Title.
QC173.57.F7513 1994
530.1′1—dc20 94-3876
CIP

♾ The paper used in this publication meets the minimum requirements of the American National Standard for Information Sciences—Permanence of Paper for Printed Library Materials, ANSI Z39.48-1984.

FOR BRIGITTE, OLIVER,
AND PATRICK

Most books about science that are intended for general readers are more concerned with impressing readers than with explaining elementary goals and methods clearly. When intelligent nontechnical readers get hold of books of that kind, they become thoroughly discouraged and conclude, "This is over my head; I can't handle it." What is more, the presentation is often sensationalized, which further repels the sensible reader. In a word, the fault lies not with readers but with authors and publishers. My advice: no book of this kind should be published until it has been determined that its contents can be understood and appreciated by an intelligent and critical general reader.

ALBERT EINSTEIN

Contents

Preface to the English Edition ix
Editorial Note x
Introduction xi

1 Newton and the Ocean of Truth 1
2 Newton and Absolute Space 10

HALLER'S DREAM

3 Meeting Newton 27
4 A Dialogue on Light 36
5 Newton Meets Einstein 53
6 The Speed of Light as a Constant of Nature 64
7 Events, World Lines, and a Paradox 72
8 Light in Space and Time 91
9 Time Dilation 100
10 Fast Muons Live Longer 120
11 The Twin Paradox 134
12 Space Contraction 145
13 The Marvel of Space-Time 154
14 Mass in Space and Time 165
15 An Equation That Changed the World 177
16 The Power of the Sun 189
17 Lightning at Alamogordo 202

18 Energy Hidden in the Nucleus 208
19 Mysterious Antimatter 222
20 Marveling at Elementary Particles 235
21 Does Matter Decay? 245

Epilogue 255
Sources of Quotations 257
Suggested Reading 259
Glossary 261
Index 267

Preface to the English Edition

On April 2, 1921, the Dutch ship *Rotterdam* steamed into New York Harbor with Albert Einstein on board. Einstein's arrival in the United States marked the beginning of an enduring fascination both with the ideas of relativity and—unprecedented for a scientist—with their creator. In 1921, however, nobody suspected that Einstein's theory concerning the relativity of space and time and the equivalence of matter and energy would play, more than twenty years later, a major if indirect role in world politics, influencing the personal lives of all members of the industrialized nations. As it has turned out, Einstein's insights, combined with recent findings in nuclear physics, particle physics, and astrophysics, have led to a new view of our physical world, which began about fifteen billion years ago in a violent explosion, the Big Bang.

Some knowledge of Einstein's ideas is essential background for anyone seeking to understand modern cosmology. Such knowledge is provided by this book, now available for English-speaking readers, as it was earlier for German and Italian readers. I should like to thank Penelope Kaiserlian of the University of Chicago Press for making publication of this edition possible; Karin Heusch for her great care with the translation; and Clemens A. Heusch for expert assistance with terminology. I hope the book will help familiarize general readers with some of the marvels of modern science. Not only is this wider knowledge crucial for the further promotion of scientific research; it is also a part of Einstein's legacy.

CERN, Geneva, April 1994

Editorial Note. Readers should note that in this translation the term *billion* is used in the American sense of one thousand million (10^9) and that all temperatures are expressed in kelvins.

Introduction

> Einstein explained his theory to me every day; by the time we arrived, I was finally convinced that he understood it.
>
> —CHAIM WEIZMANN, DESCRIBING A TRANSATLANTIC CROSSING WITH EINSTEIN IN 1921

The title of this book is unusual; it refers to a mathematical equation:

$$E = mc^2.$$

This equation describes the connection between energy E and mass m of a material object; these two quantities are connected by the speed of light c, which is about 300,000 kilometers per second. Einstein's famous equation, which he wrote down in 1905, is not merely one of those mathematical formulas that undergird modern physics; it is a very symbol of our time. This became clear, at least to the scientists and technicians involved in the first test of an atomic bomb, when their nuclear device detonated in a New Mexico desert on July 16, 1945, at six o'clock in the morning. For the rest of the world it became clear a few days later when, on August 6, 1945, more than a hundred thousand people in Hiroshima fell victim to an atomic explosion.

Since that time, the consequences of the relation between energy and mass have determined world politics either directly or indirectly in the form of atomic bombs and hydrogen bombs. The mere possibility that all life on our planet might be destroyed by these bombs has given us a long period, from 1945 to the present, without global wars. Instead, countries with nuclear weapons have been keeping each other in check, in an unstable balance.

It is too early to judge how long this equilibrium can be maintained, and whether the threat of potential global annihilation will ultimately permit universal disarmament. It is an irony of history that a world without war, that is, without atomic war, may finally become possible because we have realized that the alternative would not be the world as we have known it, where war was considered a legitimate means of international politics, but *no world at all.*

The beginning of the twentieth century was marked by worldwide political changes that led to the breakdown of the apparently well-ordered bourgeois world of the late nineteenth century. Among these are the beginnings of an organized revolutionary movement in Russia, the economic and political rise of the United States of America, and the emergence of the potential for large-scale conflicts in Europe that finally led to the outbreak of World War I. Interestingly, a revolutionary rethinking of the sciences began about the same time. This was triggered by a rather conservative German physicist, Max Planck, who laid the foundations for quantum theory and, hence, modern atomic physics, and by a young employee of the Swiss patent office in Bern, named Albert Einstein.

Toward the end of the nineteenth century, the natural sciences were dominated by classical physics, whose crowning glory was Isaac Newton's laws of mechanics. These laws were seen to be uniformly valid throughout our cosmos, governing the motions of stars, planets, and atoms. The basis of Newton's mechanics was the stability and immutability of mass. According to Newton, space and time were given, universal structures.

Einstein's relativity theory or, more precisely, his special theory of relativity had amazing consequences. (His general theory of relativity, formulated around 1915, deals mostly with problems of gravitation and will not be discussed here.) Neither space nor time is a universally applicable concept; both depend on the physical situation of the observer. And no longer can there be universal meaning to the concept of mass: mass can be transformed into energy, and vice versa.

That transformation is described by Einstein's equation $E = mc^2$. The equation says that for each unit of matter there is an enormous corresponding quantity of energy—namely, the amount arrived at by multiplying the corresponding mass by the square of the speed of light c.

How great this energy can be may be illustrated in the following example: a moving car, driven at a speed of 180 kilometers an hour—that is, 50 meters a second—has a kinetic energy of one-half times the mass m times the square of the speed v, that is, $\frac{1}{2} mv^2$. According to Einstein's formula $E = mc^2$, the energy corresponding to the mass of the car is greater by a factor of $2 \times (c/v)^2 \approx 7.2 \times 10^{13}$, that is, by a factor of almost 100 trillion (10^{14}).

This energy cannot, of course, readily be put to use, since the materials from which the car is made are stable; they cannot be transformed into other forms of energy, such as radiation energy.

That transformation is possible only with the help of nuclear physics technology, and even then it can be achieved only partially.

Einstein's equation describes not only the transformation of matter into energy but also the inverse process—the transformation of energy into matter. It is possible, for instance, to produce particles of matter through the collision of particles of light, or photons. This possibility allows physicists and astrophysicists to speculate about the production of matter at the beginning of the evolution of the cosmos, the so-called Big Bang.

There is a mistaken notion that relativity theory is too complicated, that it can be understood only by experts. This is true with respect to the details of the theory, which are certainly difficult. The basic ideas, however, are fairly straightforward, and the interested layperson should have no trouble grasping them. The problems encountered by the expert seeking to explain them to an interested audience of nonphysicists are of a conceptual nature.

From early childhood, all of us have developed a sense of the space around us and of the apparently regular and universal flow of time. Certain consequences of relativity theory are often described as if they were in contradiction to this sense. We get the false impression that relativity theory deals with a total revolution of the concepts of space and time. In reality, however, the theory is a modification and an expansion of those concepts and applies to situations that rarely, if ever, occur in our daily lives—more specifically to processes in which matter moves at a tremendous speed, close to the speed of light.

The speeds we deal with in daily life are very small by comparison. Hence, our intuitive understanding of space and time cannot accommodate the strange effects that are anticipated according to relativity theory in cases of extreme speeds. To understand such effects, we must not only learn new things but must abandon familiar ideas or see their limitations. And there lies the true difficulty.

To abandon old, sometimes centuries-old ideas is a painful process that can often be accomplished only with great effort. The secret of important discoveries in the natural sciences lies less in the generation of new ideas than in recognizing the insufficiency of familiar ones.

When Einstein, shortly after the start of the century, discovered the relation between energy and mass, he started from the idea that the equation $E = mc^2$ was simply a useful one for evaluating the equivalence of energy and mass in physical processes. In the physical processes that had been studied in detail by that time, there was no way of practically, and directly, converting mass into energy—say, into electromagnetic radiation. At best, only minute fractions of a particular mass could be converted into other forms of energy.

Einstein himself did not then believe it would ever be possible to transform large amounts of matter directly into energy. But he was mistaken. He could not have known that only a few years after he formulated his equation, new forces would be found—the strong forces inside the nucleus of the atom. With their help, a relatively large portion of the mass of an atomic nucleus can be changed directly into energy, whether the kinetic energy of particles or the electromagnetic energy of photons emitted in the process. That is what happened in the explosion of the atom bomb dropped on Hiroshima.

On August 6, 1945, about one gram of the mass of the bomb was suddenly changed into energy—an amount of energy equivalent to the explosion of 12,400 tons of the conventional explosive TNT. (The bomb itself, a complicated technical device, weighed much more than a conventional bomb—almost four tons.) It was enough to obliterate the greater part of a city with a population of about 300,000.

The energy in an atom bomb or in a nuclear reactor is produced by converting mass into energy. This works only because, alongside the forces everybody knows—those of gravitation and electromagnetism—there is a further natural force: the strong interaction between the particles inside the nucleus of an atom. (There is also another force, the so-called weak nuclear force, which manifests itself in the radioactive decay of atomic nuclei; that need not concern us here, however.) An essential feature of this strong interaction is that the objects (particles, nuclei) involved in a strong interaction process undergo frequent change in their mass.

Reactions due to the strong interaction were commonplace during the early development of the cosmos. Thus it is possible to explain the synthesis of light atomic nuclei by strong interaction processes shortly after the Big Bang—which probably occurred some 15 billion (15×10^9) years ago. Heavy nuclei, as in iron, have been synthesized during nuclear processes in the interior of stars. The energy radiated from the sun, which benefits our daily lives, is likewise produced by strong interaction processes in its interior.

Processes involving strong interaction no longer occur on Earth. Most of the dynamic processes on Earth are governed by gravitational or electromagnetic interactions. In these processes, the masses of the particles involved do not change.

Chemical processes, such as ordinary combustion or the explosion of a grenade, are ultimately electromagnetic processes, since atoms are held together by electric attraction. That is why it was long not understood that mass and energy are interchangeable. In the nineteenth century, physicists and chemists spoke of two different conservation laws in the natural sciences: conservation of energy, and conservation of mass.

It became clear only at the beginning of the twentieth century that a number of the phenomena discovered in the nineteenth, such as those involving electromagnetism or atomic processes, can be understood only if space and time are not considered separately. Space and time must be seen jointly: we use the term "space-time." Formulating this union of space and time mathematically is what Einstein's relativity theory is about. An important consequence of this union is the interchangeability of mass and energy.

In classical mechanics, which we may also call Newton's physics, mass and energy exist as two separate concepts. According to Newton, the energy of a cannonball at rest is zero; in Einstein's theory it is enormously high. Yet we shouldn't say that Newton's theory has been replaced by relativity theory. It remains valid in situations in which no speeds are comparable to the speed of light, which, as we have mentioned, is about 300,000 kilometers a second. Most of what we observe in daily life takes place in this limited situation. So Newton's physics makes immediate sense in terms of our everyday experiences: anyone who drives a car has an intuitive familiarity with these laws. In critical situations, no driver can survive without this intuition.

In processes involving change in atomic nuclei by means of strong interaction, the speeds of the particles involved are often comparable to that of light—frequently exceeding 100,000 kilometers a second. Understanding these processes necessarily involves relativity theory and, therefore, the concept of the unity of space and time, of mass and energy.

The importance of relativity theory is not limited, however, to our understanding of the workings of an atom bomb or a nuclear reactor. In recent times its effects have become important in many fields of the sciences and of technology; they are relevant to particle accelerators and to medical instruments, as well as to electronic devices. Today, the basic elements of the theory of relativity should be just as much a part of general knowledge as is the atomic structure of matter.

Numerous books have been written on relativity theory for nonphysicists, one by Einstein himself (see the Bibliography). My own presentation differs from all others in two respects.

First, I attempt to describe the far-reaching consequences of the special theory of relativity as it concerns today's ideas of the structure of matter, where the connection between mass and energy plays a central role. This emphasis on the material world and the equivalence of mass and energy is suggested by the title of the book. Anyone writing on science for the general public must carefully choose what to say and, more particularly, what not to say. I have deliberately omitted a detailed discussion of cosmology and the Big Bang. This book is a novel summary of the many aspects of Einstein's mass/energy equation, a motif that threads its way through modern physics like a red line, and that can be traced back to the beginning of the universe, the primeval explosion of matter.

The major part of this book is written in the form of a fictitious discussion between Isaac Newton, Albert Einstein, and a third, invented character, Adrian Haller, professor of theoretical physics at the University of Bern. Their conversations are pure invention, since the parties concerned obviously never met. The personae "Newton" and "Einstein," as they act and argue in this book, cannot be identical with the historical persons: I merely describe possible actions and statements of Newton's and Einstein's that we might see or hear if we could get them to express their opinions on the insights and perceptions of today's physics.

The basic principles of the special theory of relativity had been developed by about 1909. That year marked the beginning of Einstein's fame as a physicist and brought in the first offers of professorships. As he appears in the discussions with Newton in this book, Einstein should be seen as the thirty-year-old man that he actually was in 1909. At that time he was quite comfortable with the special theory, but he was unaware of the consequences it would have for nuclear physics, particle physics, cosmology, and many other fields.

Newton is depicted here as he might have appeared after the completion of his great book, the *Principia.* He was then in his early forties, in his most creative period.

I chose the dialogue form because it permits contrasting ideas to be effectively presented. The difficulties of describing the principles of relativity theory are of a conceptual nature. Thus the reader should constantly be reminded of the subtle conceptual differences between classical physics and relativity theory.

Unbiased readers may first identify with Newton. Like him they may initially be hesitant about adopting Einstein's and Haller's conclusions, yet gradually become convinced relativists, as Newton does in this book.

The dialogue form was made famous by Galileo's *Dialogue on the Two Chief World Systems,* which appeared in 1632 and was responsible for widespread acceptance in Europe of Copernicus's view of the world. Unlike Galileo, I adopted the form of informal conversations in a continuing story framework.

The first two chapters present the physical concepts of space, time, matter, and so forth, as Isaac Newton formulated them. Since these concepts are closely related to the intuitive notions we all develop in childhood, readers will have no trouble accepting them. I describe in greater detail Newton's abstract ideas of absolute space and absolute time, those concepts of classical physics that undergo fundamental revision in relativity theory.

In chapter 3 the dialogue begins that is continued through most of the book. It opens with a discussion between Haller and Newton, in Cambridge, on the need for a revision of Newton's ideas on space and time. The starting points of these deliberations are new ideas about the nature of light, which are described in chapter 4. Newton's wish to exchange ideas with Einstein, the creator of relativity theory, is fulfilled in the fifth chapter, when Haller and Newton arrive in Bern.

Newton is shocked by Einstein's remarks on the constancy of the speed of light (chapter 6). Step by step, he is introduced to the ideas of the young Einstein, as Einstein and Haller act as his guides into the world of relativity.

The first step toward relativity theory is taken in chapter 9: Newton is confronted with the phenomenon of time dilation at exceedingly high speeds. The following chapter describes its experimental verification. In chapter 10, Newton accepts the experimental proof of time dilation by observation of fast muons; and he also accepts the possibility of twins, one of whom has departed on a space voyage, aging at different rates (chapter 11).

Newton eventually accepts both the apparent shortening of a rapidly moving object (chapter 12) and the astonishing symmetry between space and time as discovered by Einstein (chapter 13). In chapter 14, Newton becomes acquainted with the new formulation of mass according to relativity theory. Mass and energy, like space and time, are now closely related.

Newton himself introduces the famous formula $E = mc^2$. From this point on, it is Haller who takes the lead in the discussions. The remaining conversations are held at CERN, in Geneva; they deal with subjects of which neither Einstein nor Newton has immediate knowledge.

Chapter 16 deals with nuclear fusion and nuclear fission. The explosion of the first atomic bomb on July 16, 1945, in the New Mexico desert, and the preparations in Los Alamos leading up to it, are the topics of the following chapter. Fission and fusion as methods of controlled energy production are discussed in chapter 18.

Chapter 19 focuses on the most impressive transmutation of mass into energy—matter becoming pure radiation on contact with antimatter. This leads to the physics of elementary particles (chapter 20) and to the cosmological question of the origin and the ultimate annihilation of all matter in the universe (chapter 21).

My principal aim in writing this book was to inform a broader public of the pivotal importance of Einstein's mass/energy equation for our understanding of the physical world. At a time when the generating of nuclear energy, both for today and for the future, is being widely discussed, interested readers without scientific expertise should be in a position to see for themselves the significance of relativity theory.

Many people see Einstein's formula as a magical code invented by physicists, not as a profound property of nature. I hope that in the not too distant future the ideas of relativity theory will shed their shroud of magic, mystery, and incomprehensibility and will become a part of our common education. And I hope this book will contribute to that goal.

I wrote part of the manuscript at CERN while a guest of the theory division there. For their hospitality, I would like to thank the members of that division. I would also like to express my gratitude to the late Richard P. Feynman of the California Institute of Technology in Pasadena for the valuable discussions I had with him about the form and topics of this book. And I am grateful to the theory group of the Los Alamos Scientific Laboratories in New Mexico for their hospitality during a summer stay which saw the conceptual beginnings of this book.

Munich, October 1992

ONE

Newton and the Ocean of Truth

In late July, just as the semester break had started at Bern University, Professor Adrian Haller flew to a meeting at the University of California in Santa Barbara. He had left Bern early so that he could visit friends in London on the way. On the day of his arrival there, he found the time to visit Isaac Newton's tomb in Westminster Abbey. Standing in front of the monument, Haller read the epitaph:

> *Sibi gratulentur mortales tale tantumque existisse humani generis decus* (Humans should be grateful that someone who so adorned their species lived among them).

The words reveal the respect and admiration in which Isaac Newton is held to this day by his British compatriots. Born in Woolsthorpe, Lincolnshire, on December 24, 1642 (according to the Julian calendar)—the year of his great Italian colleague Galileo's death—Newton died March 20, 1727, in London.

The significance of Newton's ideas for the development of our view of the world can hardly be overestimated. No other natural scientist, before or since, with the possible exception of Einstein, did so much to shape the development of the natural sciences and technology. Even poets were impressed by the clarity and keenness of his thought, witness Alexander Pope's well-known lines:

> Nature and Nature's laws lay hid in night:
> God said, let Newton be! and all was light.

Since Haller planned to spend the weekend in England, he decided to visit Newton's workplace, Trinity College in Cambridge, some fifty miles northeast of London. He arrived in Cambridge on Sunday, a beautiful summer day. After a short walk through the town, he reached Trinity College. He quickly found, to the left of the Main Gate, the small, plain building in which Newton had long lived and worked.

No one was to be seen in Trinity's great quadrangle on that Sunday morning. Haller sat down on the steps of the fountain in the middle of the quadrangle to enjoy the sun and the quiet. No

Fig. 1.1 Isaac Newton, age 46; portrait by Godfrey Kneller. This is the earliest likeness we have of the great physicist. (Reproduced with the permission of Lord Portsmouth and the Portsmouth Bequest.)

Fig. 1.2 While a fellow of Trinity College, Newton lived in the small building that connects the Main Gate of the College with the Chapel. His rooms were on the second floor next to the gate.

one disturbed him. The only person he saw was a middle-aged man, probably a scientist or a tutor at the college, sauntering through the gate and walking into Newton's abode.

Haller tried to imagine how things might have looked during Newton's time here. It probably hadn't been much different from today. The college had not much changed over the centuries. Newton arrived as a student in Cambridge in 1661 and enrolled at Trinity College. His main interests were mathematics, astronomy, chemistry, and (it should not be forgotten) theological studies. As a student he impressed Isaac Barrow, who held the Lucasian professorship in mathematics (named after Henry Lucas, who had endowed the chair). Barrow's scholarship was not limited to the natural sciences and mathematics; he also retained a lively interest in languages and in religious issues. He had been a preacher, a professor of Greek with the requisite knowledge of Latin and Hebrew as well as Arabic. He had also taught optics and mathematics.

There is no doubt that Barrow profoundly influenced the development of his young student Isaac Newton. Through him, Newton not only became acquainted with the scientific ideas of his time; it was also thanks largely to Barrow's guidance that he developed, in his later years, an uncommonly strong interest in religion. In particular, Barrow introduced Newton to the ideas of Spinoza and Hobbes.

At age twenty-three, Newton received the degree of bachelor of philosophy. He wanted to stay at Cambridge to study mathematics but had to postpone that project. In 1665 England fell victim to the black plague. The authorities closed all universities to lessen the danger of an epidemic. Newton returned to Woolsthorpe to his mother's house; yet this turned out to be Newton's most productive time. In a year and a half he developed not only the basic ideas of differential and integral calculus but also those of classical mechanics. He formulated the universal law of gravity—the general attraction between masses—that remained one of the pillars of physics until, 250 years later, Albert Einstein gave that law a fundamentally new interpretation.

The wealth of ideas Newton came up with during his stay in Woolsthorpe can be explained only by his exceptional powers of concentration. Emilio Segrè says of Newton: "His peculiar gift was the power of holding continuously in his mind a purely mental problem until he had seen straight through it. I fancy his preeminence is due to his muscles of intuition being the strongest and most enduring with which a man has ever been gifted. Anyone who has ever attempted pure scientific or philosophical thought knows how one can hold a problem momentarily in one's mind and apply all one's powers of concentration to piercing through it, and how it will dissolve and escape and you find that what you are surveying is a blank. I believe that Newton could hold a problem in his mind for hours and days and weeks until it surrendered to him its secret."

In the history of natural science, there is only one other example of an abundance of ideas coming forth in such a brief time. That was in 1904–5, when Albert Einstein worked out the basic ideas of the relativity of space and time—an important continuation of Newton's ideas. Einstein was also to become one of the founders of modern quantum theory.

After returning to Cambridge, Newton impressed Barrow so much that the professor decided to submit some of the results of Newton's research, with his permission, to members of the Royal Society. The society had been founded in London in 1660. Thus Newton's name became known for the first time outside of Cambridge. When Barrow retired from the Lucasian chair of mathematics in 1669, he must surely have been influential in the choice of his successor: the twenty-seven-year-old Isaac Newton.

Newton's first lectures dealt with the study of optics. Apart from his theoretical research, he used his rooms at Trinity to experiment with various pieces of apparatus, most of which he had built himself. Although his fame today is based on the creation of physical theories, Newton was also an excellent experimentalist and craftsman. Among the evidence for this still in existence is the mirror telescope that is preserved in the Royal Society's collection. Its mirror was polished by Newton himself.

Newton's first scientific publication, which appeared in 1672 in the *Philosophical Transactions of the Royal Society,* deals with optics, specifically the connection between the diffraction of light and its spectral color, a connection that Newton himself discovered. This discovery later proved important for the clarification of the physical nature of light.

More than two hundred years later, it was again reflections on the nature of light that would initiate a revolution in the concepts of physics, this time triggered by Albert Einstein. He gave his opinion of Newton's research in his introduction to a new edition of Newton's *Opticks*: "Lucky Newton, blessed childhood of science! He who has time and leisure can, by reading this book, relive the wondrous experiences of the great Newton in his early years. Nature to him was an open book, with letters that he could read effortlessly. The concepts he applied to bring order to observed phenomena came directly from experience; they issued from the elegant experiments he mounted, one after the other, like toys, and which he described in loving detail. In his person, he united the experimentalist, the theorist, the mechanic, and, last but not least, the showman. Strong, assured, and alone, he stands before us: His creativity and his accuracy down to the last detail are manifest in every word and in every number."

Newton published the results of his research only reluctantly. He usually waited until the danger of a priority conflict with other scientists loomed. It is to the credit of the astronomer Edmund Halley (1656–1742) that he convinced Newton to make his ideas and results known in a major publication. In 1687, Newton's magnum opus appeared, the *Philosophiae Naturalis Principia Mathematica* (Mathematical Principles of Natural Philosophy).

This book, usually referred to simply as the *Principia,* is among the keystones of the physical sciences. It laid the foundation of mechanics and thereby of the development of technology. In his introduction, Newton describes his approach to physical phenomena: "From the phenomena of motion to investigate the forces of nature, and then from these forces to demonstrate the other phenomena." In the three hundred years since the appearance of the *Principia,* we have witnessed the exceptional success of Newton's research methods.

The *Principia* are organized in three parts, preceded by Newton's famous definitions of the fundamental concepts of mechanics, which I shall discuss in more detail later.

Book I is devoted to various questions of mechanics. In particular, it focuses on the motion of rigid bodies under the influence of central forces, those forces directed toward a central point, such as the attractive forces of gravitation that originate from the sun and determine the motion of the planets.

Book II deals with applied physics. Newton investigates, among other things, the motion of rigid bodies in media such as air and water. What resistance, for example, does a body experience when it moves through such a medium? In this connection Newton established a new branch of mathematics whose importance for physics would become known only a century later, namely, variational calculus. Book II concludes with a discussion

PHILOSOPHIÆ

NATURALIS

PRINCIPIA

MATHEMATICA.

Autore *J S. NEWTON*, *Trin. Coll. Cantab. Soc.* Matheſeos Profeſſore *Lucaſiano*, & Societatis Regalis Sodali.

IMPRIMATUR.

S. PEPYS, *Reg. Soc.* PRÆSES.

Julii 5. 1686.

LONDINI,

Juſſu *Societatis Regiæ* ac Typis *Joſephi Streater*. Proſtat apud plures Bibliopolas. *Anno* MDCLXXXVII.

Fig. 1.3 Title page of Newton's most important work, the *Principia,* 1687 edition. Bancroft Library, University of California, Berkeley.

of the theory of waves, which Newton limits to the travel of sound waves and mechanical waves in water.

Book III, titled "The Systems of the World," covers astronomical phenomena. On the basis of his theory of gravitational attraction, Newton offers an explanation of the motion of the planets in the gravitational field of the sun—a scientific tour de force that accounts for his universal fame.

At the end of the *Principia,* Newton says of his theory of gravitation:

> Hitherto, we have explained the phenomena of the heavens and of our sea by the power of gravity, but have not yet as-

signed the cause of this power. This is certain, that it must proceed from a cause that penetrates to the very centers of the Sun and planets, without suffering the least diminution of its force; that operates not according to the quantity of the surfaces of the particles upon which it acts (as mechanical causes used to do), but according to the quantity of the solid matter which they contain, and propagates its virtue on all sides to immense distances, decreasing always as the inverse square of the distances. Gravitation towards the Sun is made up out of the gravitation towards the several particles of which the body of the Sun is composed; and in receding from the Sun decreases accurately as the inverse square of the distance . . . But hitherto I have not been able to discover the cause of those properties of gravity from phenomena, and I frame no hypotheses [in the Latin original, the famous expression *Hypotheses non fingo* appears here]; for whatever is not deduced from the phenomena is to be called a hypothesis; and hypotheses, whether metaphysical or physical, whether of occult qualities or mechanical, have no place in experimental philosophy. In this philosophy, particular propositions are inferred from the phenomena, and afterwards rendered general by induction. Thus it was that the impenetrability, the mobility, and the impulsive force of bodies, and the laws of motion and of gravitation, were discovered. And to us it is enough that gravity does really exist, and act according to the laws which we have explained, and abundantly serves to account for all the motion of the celestial bodies, and of our sea.

The success of Newton's mechanics, as explained in his *Principia,* was immediate both in England and on the Continent. Voltaire, for instance, alludes to Newton's ideas in many of his speeches and writings.

Newton's celestial mechanics accounted for every detail of planetary motion. It achieved a particular triumph more than a hundred years after his death, when the planet Uranus was discovered. For a while, the details of that planet's orbit had appeared to be beyond the power of his theory. Small anomalies found during precise measurement of this orbit could not be accommodated by Newton's theory of universal gravitation. One possible explanation, proposed in 1846 by Urbain J. J. Le Verrier and John Couch Adams, independently of each other, was not in con-

Fig. 1.4 Old British one-pound note bearing the image of Sir Isaac Newton, probably the most renowned Inspector of the Mint. He is depicted here together with the telescope he developed, with a prism that he was the first to use for the spectral analysis of light, and with an image of the elliptic planetary orbits around the sun. Also shown are the branches of an apple tree. An apocryphal tradition reports he conceived the idea of universal gravitation while watching an apple drop from its tree.

tradiction to Newton's theory. It assumed the presence of another planet more distant then Uranus; the gravitational influence of that other body would explain the anomalies. Le Verrier and Adams were able to indicate the exact position of the second planet, which was supposed to rotate around the sun at a radius of 4.5 million km. In the same year, the new planet was sighted by the German astronomer Johann Gottfried Galle, and was named Neptune. Here was further proof that Newton's theory could explain the most minute details of planetary motion.

Thanks, perhaps, to the success of Newton's theory of mechanics, it was a long time before its foundations were subjected to critical examination. Newton himself, no doubt, maintained a critical attitude toward the basic elements of his theory, especially the ideas of space and time. But since he always was careful in his formulations, and since in his writings, if not in his thinking, he adhered to the motto *Hypotheses non fingo,* he left no trace of such doubts.

In the course of the nineteenth century it became clear that not all physical phenomena could be explained by Newton's theory of mechanics. Certain electromagnetic phenomena simply could not be accommodated. At the end of the century, the nascent science of atomic physics proved inexplicable in terms of mechanical models, as did certain curious properties of corpuscular and gaseous matter.

Almost 220 years after the publication of the *Principia,* Newton's view of the world was finally shaken to its roots. In 1905, a twenty-six-year-old employee of the patent office in Bern named Albert Einstein published his new ideas of the inner structure of space and time. These ideas amounted to a revolutionary reshaping of the basis of mechanics. While Newton's physics was shown not to be wrong and, in many instances, to come close to reality, it now came to be seen as merely a first approximation of Einstein's mechanics.

After the original publication of his *Principia,* Newton's fame spread across Europe; he was soon acclaimed as the greatest living scientist. In 1696 the king of England appointed him Inspector of the Mint, an office of great importance. (Newton was responsible for the reform of the English monetary system.) Later he became its director. In that capacity, he is shown on the one-pound note as Sir Isaac Newton—he was knighted by Queen Anne in 1705 for his services to the Mint.

In the last twenty-four years of his life, Newton was president of the Royal Society. This society, the oldest scientific association in England, began informally in 1645 as weekly gatherings of generally recognized philosophers and natural scientists. It was formally acknowledged by King Charles II in 1660. Newton ruled the society, and thereby all scientific life in England, with an iron hand. No new member could be added without his consent.

TWO

Newton and Absolute Space

Any discussion of Einstein's ideas of the relativity of space and time should start out from a clear picture of Newton's ideas. We shall therefore review these before giving Newton and Einstein themselves a chance to state their positions.

The first concept Newton introduces in his *Principia* is that of a rigid body, or particle. He explains its mass as a product of density and volume. This definition may appear tautological as long as density has not been properly defined, and it was rejected by many critics as a mock definition. Though not unjustified, this reproach fails to recognize that Newton's ideas on the structure of matter begin with the concept of atoms.

Matter, in Newton's opinion, consists of very small particles, or atoms. The density of matter is merely a measure of the number of material particles per unit volume. This concept was essentially confirmed in the nineteenth century by the development of atomic theory.

Newton had obviously spent a long time considering how mass should be defined. Today such reflection appears well justified: despite the many insights that we have gained over the past three centuries, including quantum mechanics and particle physics, it is still not clear what exactly is meant by mass and matter.

Newton was the first to recognize the importance of the product of a body's mass and speed, which we call its momentum. This quantity doesn't change, that is, it remains constant, in the absence of an external force. Since the mass of a body, under such circumstances, generally doesn't change, its speed doesn't change either. Newton states in his *Principia* that every rigid body remains in a state of rest or of uniform rectilinear motion as long as no external forces act upon it. In our time, when we all experience motion at relatively high speeds, that statement seems quite understandable. During Newton's lifetime, it was not so. For a long time it was believed that all motions were directly related to forces.

The world around us, as we see it with our own eyes, presents itself in many facets. We observe a multitude of happenings as diverse as the falling of a leaf in autumn and the flight pattern of a bird above city roofs. All these phenomena have one thing in common: they are due to the simultaneous action of many different processes. Leaves detach themselves from trees because a slight wind moves them. They fall slowly—not as fast as an apple would fall from a branch—because the air inhibits their motion. Air resistance is present also in the fall of the apple, but it has less of an effect there.

What is the reason for these different motion processes that we observe in nature? What is motion itself? Intuitively, we would think that motion originates from a force. Take, for instance, a car, initially at rest. To put it in motion we have to exert force—say, by pushing it from the rear. To keep it in motion, we must either keep pushing or start the motor so that it takes over our job. We have the impression that motion is a state that constantly needs the action of force, or energy. This principle was formulated two thousand years ago by Aristotle, when he said that every moving body comes to rest when the force that causes the motion ceases its action.

Aristotle's principle is certainly correct; we constantly observe the ceasing of motion. But his principle has one important disadvantage—it is hardly applicable in the form in which he gave it. Aristotle, of course, was thinking of the motion of a body on Earth, where each body is in constant contact with its surroundings. His principle surely doesn't concern heavenly bodies moving in space. A spacecraft moving far away from stars or planets needs no force to maintain its motion. It will never stop but will continue in motion indefinitely.

Galileo was the first to point out that Aristotle's principle needed to be replaced. By means of many experiments, Galileo discovered that a body not under the influence of an external force maintains its motion uniformly along a straight line.

In no way is speed, as we can see, a measure for the force being exerted. If it were, a car with the engine turned off, moving on a straight stretch of freeway at 100 kilometers an hour, should remain in this state of motion, provided no external force acts upon it. In reality, that will not happen. The car will stop after a few minutes because of continuous energy losses due to the friction of the tires on the road surface and air resistance. In this sense, the car follows Aristotle's principle. So we can't say that Aristotle's principle contradicts Galileo's and is, in fact, wrong. Rather, the principle as formulated by Aristotle is insufficiently clear and hence useless for many applications, especially in technology.

Galileo's principle plays a fundamental role in Newton's *Principia.* Newton elevated it to the first rank among his laws of the motion of rigid bodies, whose tendency to remain in uniform motion he named the principle of inertia.

Newton's definitions of the concepts of space and time were of central importance for the task he set himself—to formulate the basic laws of mechanics. All things move in space and time. But what *is* space and what *is* time? Is space infinitely great, or does it have boundaries? What causes the constant flow of time? What *is* time?

Saint Augustine answered that question: "I know what time is, but if someone asks me, I cannot tell them." Thomas Mann, in *The Magic Mountain,* asked:

> What is time? Time is a mystery—unreal but still almighty. It is a condition of the world such as it manifests itself; it is motion coupled to and intermingled with the existence of matter in space and with its motion. Would there be no time if there were no motion? No motion if there were no time? Just ask. Is time a function of space? Or is it the other way around? Or are they identical? Keep asking. Time acts, time speaks, time "times." What does it time? It "times" change. The now is not the past, the here is not the there, for there is motion between the two. But the motion that we use to measure time runs in circles, is closed up in itself, and therefore it is a motion and change that can just as well be called rest and standstill. For the past constantly repeats itself in the now, the there in the here.

It is hard indeed to determine the essence of the various phenomena we call time; to this day, physicists have not fully succeeded. It is much easier for a physicist to explain how to measure time—with a clock, of course. There is the essential fact that nature provides us with periodic motions—motions that keep repeating themselves, such as the back and forth of a pendulum or the oscillations of a quartz crystal. To make a clock, all we need is a device that counts these motions. The number of cycles is a measure for how much time has passed.

Newton clearly tried very hard to establish the concepts of space and time as accurately as possible. According to him, space and time exist separately and are independent of matter. He firmly differentiates relative space and relative time, on the one hand, from absolute space and absolute time on the other.

> Absolute, true, and mathematical time, of itself, and from its own nature, flows equably without relation to anything external, and by another name is called duration: Relative, apparent, and common time, is some sensible and external (whether accurate or unequable) measure of duration by the means of motion, which is commonly used instead of true time; such as an hour, a day, a month, a year.
>
> Absolute space, in its own nature, without relation to anything external, remains always similar and immovable. Relative space is some movable dimension or measure of the absolute spaces; which our senses determine by its position to bodies; and which is commonly taken for immovable space.

It is interesting that Newton found it necessary to make a clear distinction between relative and absolute space. We all know what he meant by relative space. It is the space around us, in which we move. It gives us three different directions of motion: up and down, back and forth, right and left. In other words: our space has three dimensions. Every position in space is characterized by three numbers that are independent of each other, three coordinates. They are part of a coordinate system, which we can establish arbitrarily. The most frequently used system is defined in terms of three axes that are at right angles to each other (see fig. 2.1).

There is, of course, no absolute significance to the coordinates of a point in space. They depend not only on its actual position but also on the arbitrary choices of the origin of the coordinate system and of the direction of its axes in space. What really count are the coordinates of one point in relation to the coordinates of other points.

The three points A, B, and C, seen in figure 2.2, lie on a straight line; B is the midpoint between A and C. The property of point B that it denotes the midpoint of the path from A to C is important; it is independent of the coordinate system chosen.

Let us now consider the coordinates of these points. The differences between the coordinates of points A and B equal those between the coordinates of points B and C. If, for example, points C and A have the y coordinates 7 and 3, respectively, then point B will have a y coordinate of 5—exactly halfway between 3 and 7. The same holds for the x and z coordinates.

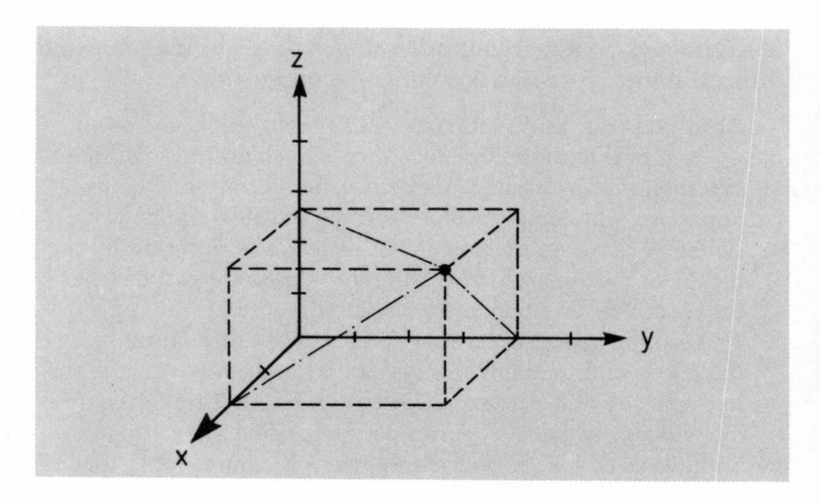

Fig. 2.1 A Cartesian coordinate system with three axes *x, y,* and *z.* The position of any point in space can be uniquely defined by three numbers, its coordinates, which are given by the projections of the point onto the three axes. The broken lines indicate these projections, which always meet the axis in question at a right angle.

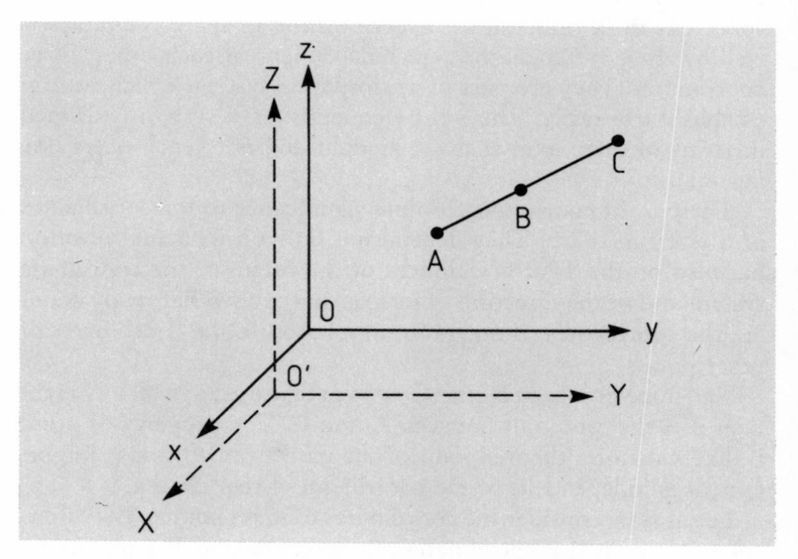

Fig. 2.2 Three points A, B, and C, shown in a three-dimensional Cartesian coordinate system. The definitions of the three points in terms of the original (full lines) and translated (broken lines) coordinate systems are fully equivalent because of the homogeneity of space.

Figure 2.2 also shows what happens when we move the origin of the coordinate system—where, for simplicity's sake, we leave the new coordinate axes parallel to the old ones. The three points A, B, and C do not change their positions; but the new coordinate system denotes them differently from the old one. Their *y* coordinates are no longer 3, 5, and 7; they may have moved to, say, 4, 6, and 8. Again, the coordinate of point B is exactly at the midpoint between those of A and C.

The property of B that it is exactly at the midpoint between A and C is independent of the coordinate system; it is said to be an invariant property with respect to a change of coordinate systems. Similarly, it doesn't matter where in space we put the three points—close to the origin or far away from it. No points or regions in space have special privilege; all have equal weight. This democratic quality of space is called the homogeneity of space. According to Newton, our space is homogeneous and infinite.

Another feature of space is important in this connection: not only can we move the origin of our coordinate system in any direction, we can also rotate the system. The directions of the coordinate axes are not fixed, because there are no privileged directions in space. Every direction is equivalent to any other—space is said to be isotropic.

The homogeneity and isotropy of space allow geometrical or physical systems—say, a triangle or a solid sphere—to be described in terms of an infinite number of coordinate systems. All are perfectly equivalent.

Because of the homogeneity and isotropy of space, we can alter the position of an arbitrary object in space without the object itself being changed. This is true as long as no external influence disturbs the homogeneity or isotropy of space. Strictly speaking, this condition applies only in outer space, far from the disturbing gravitational fields of the planets and the sun. The space in which we live our everyday lives is not isotropic. There is one direction distinguished from all others; it points toward the ground in the direction of gravity.

So far we have considered only space. But all processes in nature occur in space and time. Let us look at the simplest imaginable dynamic process in nature—the free motion of an object such as a spacecraft through space. For simplicity's sake, we shall ignore the dimensions of the spacecraft and consider it a pointlike object. This point will have a mass, the mass of the spacecraft.

For the purposes of illustration, we may call this idealized object (which does not actually exist) a "massive point."

Let us consider the free motion of this "massive point" in space. According to Newton's principle of inertia, the object either moves with a constant speed along a straight line or is at rest. In the latter case, its position is easily described in a coordinate system. The spacecraft is at all times located at a fixed point, that of its position. We can shift the coordinate system in such a way that its origin coincides with the position of a spacecraft. Then the explanation becomes very easy—the coordinates of the spacecraft are zero for all times.

It's more difficult when the spacecraft is moving in space. At any time we can denote its coordinates, but they change from one moment to the next. They depend on time. Looking at all the coordinates assumed by the spacecraft during its motion through space, we observe a straight line. At any point in time the spacecraft is located at a given place on this straight line. The line can go in any direction, and the origin may be at any distance from the line. Since space is homogeneous, we can choose to locate the origin on the straight line along which the spacecraft moves.

Fixing the straight path of the spacecraft doesn't define the dynamic sequence of its motion unequivocally. The spacecraft can move rapidly or slowly along the straight line. To determine the motion, we must establish not only *where* the spacecraft is located but also *when* it is there. We can do this easily by assigning a time corresponding to the passage of the spacecraft at each point along its straight path.

Figure 2.3 indicates times, in terms of seconds, for a specific motion. With the help of the times listed, we can determine the speed of the spacecraft. The time origin was chosen such that time is at point zero when the spacecraft passes through the origin.

The reader will have noticed that we chose arbitrarily not only the origins of the space coordinates but also the origin of our time. This arbitrariness is another freedom we have when describing a simple motion such as that of a spacecraft.

As Newton emphasized in his *Principia,* it is not only space that has a homogeneous structure; the same holds for the passing of time. For the motion of the spacecraft, the origin we choose for our coordinate system is irrelevant; neither does it matter which moment we choose as the zero point on our time scale, and whether we measure it in seconds or any other unit. The arbitrary fixing of the zero point in time by the observer of the

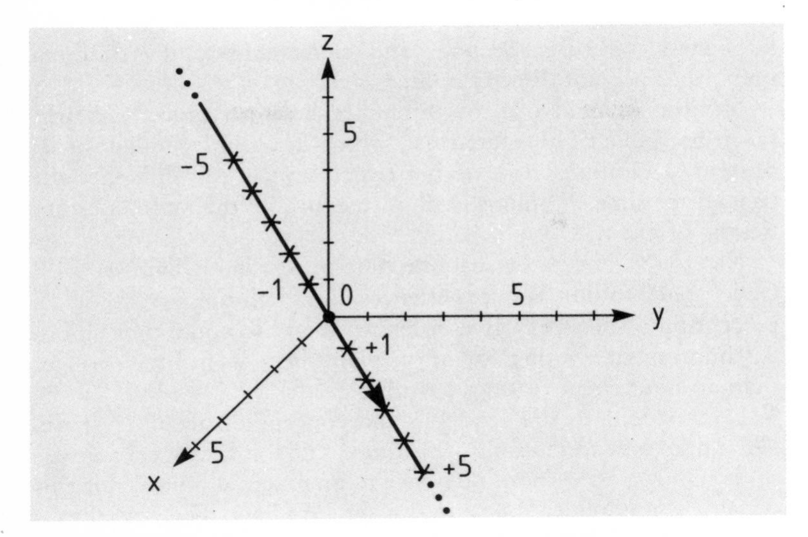

Fig. 2.3 Straight-line motion of a point mass in three-dimensional space. In the case shown here, the line passes through the origin of the coordinate system. Numbers along the trajectory mark the times, in seconds, at which the mass moved through this location. Units along the coordinate axis may be kilometers, miles, or any other measure of distance. The system was chosen such that we arbitrarily let the point mass move through the origin at time zero.

motion makes sense only if there is nothing to distinguish a particular point in time, that is, if all points of time are equivalent.

This is certainly the case when we describe the motion of a spacecraft in space. Whether the spacecraft moves along its trajectory on this day or that, or in this year or that, its motion will be the same as long as its speed is the same and as long as it moves along the same straight line.

It now becomes clear that space and time have something in common. At the very least, both are homogeneous. Time passes evenly; no point of time is basically distinct from any other. Similarly, it is immaterial which point in space we choose as point zero. There are, of course, considerable differences between space and time. In space we can move at will from one place to another. We can't do so in time, because it ticks away beyond our control. Time runs in only one direction, into the future; a trip into the past is possible only in fantasy.

There are three directions in space but only one in time. Time is indicated by the ticking of a clock; it is measured in terms of seconds, minutes, and so on. Space is measured in terms of meters, kilometers, or miles. Seconds and meters are totally different units; they are not directly related in any way.

Newton pointed out the difference between space and time when he spoke of absolute time, which flows independent of its observers. Nothing, no external circumstance, can influence the passage of time. Nothing is more inexorable than the constant ticking of a clock.

Newton's concept of absolute time makes immediate sense. It corresponds to our daily experiences, although our psychological perception of time may give different weight to equal time spans. An hour spent waiting for an airplane may seem much longer than an hour spent reading a thriller.

Less evident to the reader is the concept of absolute space. What does Newton mean by this term? "Absolute space, by virtue of its nature and without reference to an external object, remains equal and immobile," he writes in the *Principia*. The assertion is surprising: in his formulation of his principle of inertia, Newton had clearly recognized that it makes no difference whether an object is at rest or moves uniformly at a given speed along a straight line. But how can Newton speak of an absolute space that does not move?

Thus far, when I have spoken of a spatial coordinate system, I have referred to a system at rest with respect to the observer, who defines it for his or her own purposes. I should now introduce a system that moves with respect to its observer.

Let us look, for instance, at two spacecraft that meet by accident in space. Both are moving uniformly along straight lines. Inside each spacecraft is an observer who defines a coordinate system that moves along with him. Let us call these systems A and B. Now imagine we are aboard one of the spacecraft, the one with system A. Let us consider it to be at rest.

The other spacecraft passes at a speed of, let's say, 100 meters a second. Which system is better, system A at rest, or moving system B?

Newton's principle of inertia gives the answer: both are equally good. Since an absolute motion cannot be established, the observer in the spacecraft with system B will define that system to be the one at rest. With respect to the observer's system B, system A is moving at exactly the same speed as the spacecraft with system B is moving in system A. This is the speed of 100 meters a second we assumed above. Only the direction of its speed is exactly reversed.

On Earth we know of similar situations. For instance, take two trains passing each other, or one train moving past a train that is at rest in a station. Some readers will remember looking out of the window of the stationary train and having the impression that their train is actually moving, while the passing train seems to be standing still.

On Earth it is also impossible to define absolute motion. Only relative motion—motion referring to a specific coordinate system—can be defined.

Let us look at a train moving at a speed of 100 kilometers an hour. Passenger A is seated in the front car of the train, passenger B in the rear car. Both are reading a newspaper; both are at rest with respect to the train. But to an observer standing in the station motionless with respect to the surface of the Earth, both A and B are moving at the same speed, 100 kilometers an hour.

At noon, the intercom announces that the dining car in the middle of the train is open. Both our passengers get up to go to the dining car. Each walks at a speed of 4 kilometers an hour through the train, toward its middle. For another passenger on the train, both A and B are moving at the same speed, 4 kilometers an hour, but in opposite directions—A in the direction opposite to the train's motion, B in the same direction as the train.

Something totally different is being observed by the bystander at the station. B rushes by at 104 kilometers an hour—the sum of the speeds of the train's motion and of B's walking. But passenger A moves past the observer in the station at only 96 kilometers an hour—the walker's speed subtracted from the train's.

On Earth, a coordinate system that is at rest is easy to define. When we say an object is at rest, we imply it is at rest with respect to the surface of the Earth, that it isn't moving with respect to the floor under our feet or to a tree we are standing next to.

But even this is a relative statement. A passenger sitting in a moving train and reading a paper looks to his neighbor like somebody at rest, but the conductor walking through the train appears to be moving. The description obviously depends on the system of reference. For example, both the passenger and the conductor are moving with respect to the ground or to any other system—with the exception of the coordinate system that is moving along with the train.

A coordinate system defined by an observer aboard a spacecraft that is freely moving in space is called an inertial system. In an inertial system, the free motion of an object can easily be described. According to Newton's law of inertia, the object will move in a straight line unless at rest. If at rest, it will remain so for all time.

Like many other concepts in the sciences, the concept of an inertial system represents an idealized, extreme case. There is really no such thing as a spacecraft moving freely through the universe, independent of and uninfluenced by other bodies, for even if it were to move far away from planets, stars, and galaxies, it would still be influenced by the gravity of distant celestial bodies.

It is possible to realize an inertial system only approximately. For example, a coordinate system with the sun as its origin that can describe the movement of the planets is a good approximation to an inertial system. True, the sun itself is not free of outside influence; the gravitational attraction of the other stars in our galaxy forces it to move in an almost circular orbit around the center of the galaxy. The curvature of its orbit is so slight that it can be ignored for most purposes; we can safely regard this orbit as a straight line. With this slight reservation, we can consider the system with the sun as its origin an inertial system.

Things are less obvious when we look at a coordinate system that is connected with the Earth's surface. Our planet moves around the sun by virtue of the gravitational attraction of the sun. The Earth also rotates on its own axis once every twenty-four hours. Thus a coordinate system nailed to the surface of the Earth describes a fairly complicated pattern of motion when viewed from outside the solar system. It doesn't move uniformly as it would if it were an inertial system.

Strictly speaking, there are no inertial systems on Earth. We can, however, disregard the vagaries of an Earthbound system in many applications, as, for instance, in the dynamics of a moving automobile. The fact that the automobile, along with all objects on the surface of the Earth, is moving around the sun and around the Earth's axis is of little importance. If it moves uniformly and in a straight line on the freeway, we can—for most purposes—interpret the coordinate system defined by the car as an inertial system.

Traditionally, on Earth, we use a system that is at rest with respect to the Earth's surface. This too approximates an inertial system, like all other systems that move uniformly and in a straight line relative to the system at rest. The latter is the system with respect to which we define our speed. A driver who is caught on the freeway going 130 kilometers an hour in a 100-kilometer-an-hour zone will not be able to escape the fine by pleading that the speed limit makes sense only when the reference system is clearly defined.

Yet specifying speed makes sense only if the appropriate reference system is clearly specified. Speed is necessarily relative to a given system. As a result, we cannot directly sense the speed at which we are moving. Sitting in a car with our eyes closed, we cannot differentiate between speeds of 100 and 150 kilometers an hour.

Not all physical quantities related to speed are relative, however. Acceleration, for instance, is immediately felt. Sitting with eyes closed in a car, we sense immediately when the car changes speed, when it accelerates or slows down. When the car speeds up, we are pressed back into our seat: as a consequence of acceleration we become aware of a force. This force appears because every object wants to remain fixed in its state of motion. If hindered from doing so, it will oppose the acceleration with a force that is known as the force of inertia.

If we now take the moving and accelerating car as our reference system, we notice that objects formerly at rest, such as pencils lying on the dashboard, will not remain in place. Depending on the amount of acceleration, they will slide backwards.

The appearance of inertial *forces* makes it clear that we are not dealing with an inertial *system*. In accelerated reference systems, an object will generally not move in a straight line; it will tend to move along fairly complicated curved paths.

The inertial forces resulting from an acceleration can be measured and therefore have a meaning that is independent of the reference system. The acceleration of an object has an absolute meaning, unlike the object's speed, which is relative and thus dependent on the reference system.

Let us now return to Newton's idea of absolute space. We have just seen that for every inertial system there is an infinite variety of other inertial systems moving uniformly in a straight line in relation to the first one. Which of these can be said to describe absolute space in Newton's sense? Does it makes sense at all to speak of absolute space as Newton did, independent of matter and not influenced by it? Is there space without matter? Or isn't it, rather, that matter is responsible for the existence of space, since space manifests itself through the matter in it, through the many ways material objects can be arranged in it.

For Newton, the idea of absolute space had an almost mystic, even a religious meaning; to him it represented an all-fulfilling spirit, comparable to God. It is only to God that we can attribute qualities such as absolute space. Absolute space is eternal, infinite, immobile. It can be neither destroyed nor created. It is all-present. Newton presumed the creator of the universe to be an expert in geometry.

Of course, Newton saw the difficulties with which he was confronted when introducing the idea of absolute space. Its existence would imply that out of infinitely many inertial systems there is one that has a special meaning—the one system that is at rest with respect to absolute space. But this property is not measurable by experiment, as Newton freely admitted.

One could avoid these difficulties by renouncing the concept of absolute space or, as a compromise, by regarding the totality of all inertial systems as absolute space. If one were to know one of these systems, one could then construct all the others simply by taking all possible uniform and rectilinear motions into account. Apparently Newton didn't want to admit such a compromise, for he clung to his notion of absolute space. We don't know why he did this, but he was likely motivated by considerations beyond the arena of science.

Adrian Haller sat pensively by the fountain in the courtyard of Trinity College. "How strange that Newton was so rigidly fixed on the idea of absolute space," he thought. "Obviously he was abandoning his motto *Hypotheses no fingo* (I do not make up hypotheses) when he hypothesized absolute space. After all, to introduce the idea of absolute space and, along with it, an absolutely given coordinate system—hence a privileged inertial system—without telling us how to determine that system experimentally: well, that's a pretty bold hypothesis!"

At that moment, Haller was sorry not to be able to ask Newton directly. He was only a few feet from Newton's workplace, but of course it was impossible simply to walk into his rooms and ask him. Haller resigned himself to this reality, and a short time later he left the quadrangle. It was a beautiful summer morning, exceptionally warm for the English climate. The walk through the slowly awakening city of Cambridge was tiring, and Haller sat down to rest on the well-tended lawn in the park behind the college. He soon fell into a deep sleep.

But sleep did not chase away those new impressions he had just gained at Trinity College—quite the contrary. Isaac Newton was to play a special part during Haller's slumber. When Professor Haller met me a few days later at a conference in Santa Barbara, he told me all about it during a trip to El Capitan State Park, on the shores of the Pacific Ocean. In what follows, I attempt to convey the dream as he described it, with Haller himself as both narrator and participant.

Haller's Dream

CHARACTERS

ISAAC NEWTON
Professor of Natural Philosophy, Cambridge University

ALBERT EINSTEIN
clerk at the Swiss patent office in Bern

ADRIAN HALLER
Professor of Theoretical Physics, Bern University

SCENES

Cambridge, England

Bern, Switzerland, where the equation
$E = mc^2$ was first formulated

CERN (Research Center for Elementary Particle Physics),
near Geneva, Switzerland

THREE

Meeting Newton

Soon after resting in the park I returned to Trinity College; rather, I ran back as though I had an important appointment. Something was drawing me back to Newton's workplace; I didn't know what. Once again I reached the Main Gate of the college. Once again I met the middle-aged man, who was in no hurry this time; he stopped and looked at me curiously.

"I saw you here a little while ago," he said. "Are you looking for something in particular? May I help?"

I couldn't help smiling at the thought that I was actually looking for Newton, and that the search was utterly crazy—who in his right mind would look for a scientist who died 250 years ago? I answered quickly, "I'm not looking for anything in particular. I would just like to look around. I've always wanted to see the place where Newton did his work, and today I finally have the chance."

"I assume you are a physicist," said the man, looking at me closely.

"You are right. And you won't believe it, but I really am looking for something, or, rather, for someone; I'm looking for Isaac Newton."

I was surprised at my own answer, which was as honest as it was absurd; and I was also surprised that my questioner took it as something quite normal. He smiled and remarked blandly, "You needn't look any further. *I'm* Isaac Newton."

At that moment I recognized him. There he was, standing in front of me: Newton, a forty-year-old man at the time he wrote the *Principia;* the Newton I knew from Godfrey Kneller's portrait. His hair was shorter, he was not wearing a wig, and he was wearing modern clothes. An unsuspecting observer would have taken him for a twentieth-century Trinity don.

At the same moment I wondered about myself: I was taking for granted the sudden appearance of one of the greatest scientists that had ever existed. What is more, I was acting as if it were not Newton but a run-of-the-mill academic standing there in front of me. So I introduced myself: "Adrian Haller, professor of physics at the University of Bern."

Newton seemed delighted to meet a colleague from the Continent and continued, "I think I owe you an apology. You may be astonished to meet me here at Trinity. After all, it's been almost three hundred years since I worked here."

I nodded and acted as if it were the most natural thing on Earth to find Isaac Newton at Cambridge. I was surprised that he was so friendly toward me, even outgoing. In his own time he had a reputation for being reclusive, barely accessible even to his colleagues. Apparently he had changed since then, and not to his disadvantage.

"A few days ago, I was granted the opportunity to visit my old workplace," said Newton. "I've been staying in Cambridge since then. As you can imagine, a lot of things are new to me—the traffic in the streets, the bright lights in the college rooms—you modern people call it electric light, I believe. Then there are those strange moving picture devices with a curved glass pane in front, which you call television sets. I've grown used to things now. I spend most of my time in the library, trying to find out what has happened to the natural sciences that I used to work with. I must admit that I'm having serious problems. Much of what I read in modern physics textbooks I don't understand."

"No wonder," I broke in. "In the three centuries since your *Principia* appeared, lots of things have happened in science, and physics is no exception. New phenomena in atomic physics discovered at the end of the nineteenth century could no longer be understood by means of your mechanics. New theories had to be developed, specifically the theory of relativity and the theory of quantum mechanics."

"There it is again, that concept you call relativity theory," exclaimed Newton. "I've run into it several times while reading the books. What on Earth is this theory? Should it be taken seriously? Maybe you can tell me more about it? Only last night I tried to make sense of it from the few remarks I found in the textbooks, but I'm afraid I had no luck. One thing was clear to me, though—this strange theory doesn't accept my concept of absolute space."

What was I to answer, I, a physicist at the end of the twentieth century, who confronts the problems of his science every day? How could I refuse Newton's request? After all, it would be a unique opportunity to get to know the trains of thought and the capabilities of this singular genius.

"Agreed," I said. "I suggest that together we look at the most important aspects of relativity theory. But you must grant me one thing. You worked on the fundamentals of mechanics, the mainspring of technology; I, too, would like to take advantage of our meeting by gaining some access to your world of thought. I suggest we proceed systematically and look at the physical phenomena connected with relativity theory that were discovered in the course of this twentieth century. But this is going to take some time—maybe several days. How much time can you spare?"

"I'm not short of time," replied Newton. "My only hesitation is that I might take up too much of yours. When I did my research here in Cambridge, three hundred years ago, I had plenty of time. A few days were nothing at all. Things seem to have changed, though. I have the impression that hardly anybody has enough time these days to think about anything in a more than superficial way."

"As a matter of fact, I'm on my way to a conference in America," I said. "But if I have an opportunity to spend a few days with Isaac Newton, I'll leave the conference to my colleagues and stay right here."

Newton was delighted to hear more about what he called, with all due modesty, the continued development of his natural science. He agreed to set aside the next few days so that we could discuss it.

Neither Newton nor I had made plans for the remainder of the morning, and so we began our talks then and there. We decided that I would start by giving Newton today's view on the foundations of mechanics—which amounted to carrying coals to Newcastle, since our view is still based on Newton's own contribution. So I spent little time on it, and moved on to a discussion of the concept of absolute space. In due course, Newton summed up my arguments.

NEWTON

If I understand you correctly, you maintain that it is possible to introduce a coordinate system at some arbitrary point in space, at any time, and call it an inertial system. This system is defined by its property of containing the trajectory of a massive object in the form of a straight line, just as any object not subject to external forces should move. If we define one such system, we can infer an infinity of additional systems; they will move with respect to the first, uniformly, along straight lines, and with arbitrary speeds.

On the other hand, if a second coordinate system is rotating with respect to the first, the second system is not an inertial one. An object moving freely in this second system will follow not a straight line but a curved trajectory. That trajectory is produced by the rotation of the system. Here is the basic difference between inertial systems and rotating systems. I mentioned that in detail in my *Principia.* A similar argument applies to reference systems that are accelerated in an arbitrary direction.

I was always convinced that the kind of inertial force we experience when a carriage, or, if you will, an automobile, is suddenly decelerated must be linked to the structure of space. Speeds are always relative, but accelerations are absolute and don't depend on a reference system. But how can the acceleration of a body be established as absolute if there is no space that can provide the absolute framework?

Similarly, there must be something that distinguishes a rotating system from an inertial system. In my opinion, the difference lies in the structure of space itself. Space, and time too, are what are left whenever matter is removed from space.

HALLER

Are you sure that *anything* is left when matter is removed from space?

NEWTON

Of course, at least two things are left—space and time. [Clearly, Newton was not going to tolerate any opposition to his ideas of space and time.] In the beginning, God created the world, and he did so in space and in time; more precisely, he did so in absolute space and in absolute time. For me, both space and time are absolute, things of perfection, things divine. They are as absolute as God himself. We, on the other hand, don't exist in free space; restricted as we are to the rotating planet we call Earth, we can never fully experience absolute space—space defined by a coordinate system that means absolute rest. We can only approximate it.

Time as measured by clocks is by no means absolute time.

Our ways of measuring time describe it only incompletely. In the *Principia* I wrote: "Absolute space remains unchanged and motionless by its very nature, and does not depend on some external object. Relative space exists in relation to the former, which our perception defines in terms of its position with respect to other objects, and which we usually take for space immobile; relative space may be in motion when compared to it." In my opinion, the fact that all objects move in space, but that space remains uninfluenced and unchanged, gives proof positive for absolute space. It is the scaffolding into which God has embedded all matter.

Similar arguments can be made about time. Each of us has a different notion of how time passes—it may fly like an arrow, or crawl like a worm. Fortunately, a clock helps us avoid such fluctuations. Still, however we measure it, we shall get only a faint image of absolute, true time. Absolute time, by its very nature, flows evenly; it has no relation to external objects or to whatever matter there is in space.

Time is like the flowing waters of a wide river; individual events are like splinters of wood on its surface, propelled along by the current whenever they appear. The river of absolute time moves inexorably onward; it changes the future into the present, then discards it into the past.

As a young man, I spent hours in deep thought about time. There is nothing as simple or as complicated as time. Everybody lives through time, feels how time vanishes. Our existence is unthinkable outside time. But if you ask what the essence of time really is, nobody will give a satisfactory answer. Maybe there is no such answer.

I think that time, like space, is a framework into which God embeds matter. Stars and planets may pass, but time remains. It knows no beginning and no end. All through the universe, right here on Earth or in some distant galaxy, absolute time remains the same. God gave it to us as a link that ties us to the most distant reaches of space.

Newton spoke the last words almost imploringly. I had the impression that he was trying to convince himself rather than me. While engaging in this discussion, we were walking back and forth in the college quadrangle, and finally we sat down on the steps of the fountain. My interlocutor must have realized that I was listening to him skeptically.

NEWTON (with a hint of irritation)

You doubt me? I'll admit it was only after a struggle that I decided to define space and time simply by reference to absolute space and absolute time. We circumvent a lot of difficulties in this way. But it's merely a hypothesis, I'm afraid. I can't prove it. Still, it appears that success finally has proved me right.

HALLER

Yes and no. As far as I'm concerned, I can live with your mechanics without postulating absolute space.

NEWTON

But how are you to understand the difference between the inertial system and a rotating frame of reference?

HALLER

This is how I explain it to my students in Bern: If I am in a rotating frame of reference, say, a rotating disk, I can easily establish the rotation: to a certain extent, I see my surroundings rotating about me.

If I'm moving through outer space far away from Earth, I can do the same thing. If my spacecraft rotates, I will see the firmament rotate around me; so I realize, "Hey, my reference system rotates, I'm not in an inertial system." In this instance, the firmament gives the observer the means to tell the difference between rotation and nonrotation. Analogously, we can establish the rotation of the Earth without even mentioning absolute space.

NEWTON

Basically I agree with you. It is possible that absolute space has something to do with the fixed stars and with the vast masses in outer space.

Isn't it strange that the fixed stars are, in fact, fixed, thus apparently immobile, in their celestial configurations? But then, why not? In reality, some of these stars are in fairly rapid motion, but we don't perceive it because they are too far away. But even for a very distant star, we would be able to notice its motion in the night sky with our bare eyes if it were moving fast enough. So the stars' apparent immobility could be understood only if there were a basic principle that imposed a limit on the speed at which they could move, no matter how distant they were. Only a principle of that kind would ensure the apparent lack of motion.

Fig. 3.1 The center of the Coma cluster of galaxies. Only a small fraction of its more than two thousand galaxies is seen here. Their distance from Earth, as seen in the direction of the Coma Berenices constellation, is about five hundred light years. In principle, these galaxies could move rapidly past each other, like mosquitoes in a swarm; in reality, their relative motion is moderate. We can therefore define a coordinate system in which the galaxies are, on the average, at rest. Such a system would qualify as an inertial one. Newton might interpret this particular system as his "absolute space" if he were the inhabitant of, say, a planet in one of the Coma cluster galaxies.

Is there such a principle? If so, I'm prepared to give up on absolute space and to propose the principle that the stars are approximately at rest.

I was puzzled that Newton, after his long speech, would so quickly abandon his idea of absolute space. I answered: "You are right. It is amazing that the stars, or rather the distant star systems . . ."

Fig. 3.2 Panoramic view of our Milky Way (composite view by K. Lundmark, prepared from detailed photographic records). On the left, the galactic region that is located in the constellations Auriga, Perseus, Cassiopeia, and Cygnus. The center of the galaxy is seen from Earth in the Sagittarius constellation. On the right, the galaxy extends through Centaurus, Crux, Carina, Puppis, and Canis Major, visible only from the southern hemisphere. On the lower right, we see the two Magellanic Clouds; these small galaxies are satellites to ours. In February 1987 a supernova exploded in the Large Magellanic Cloud.

The galactic coordinate system is defined such that its origin coincides with the center of the galaxy. (Reproduced with permission of Lund Observatory, Lund, Sweden.)

"You mean the galaxies?" interrupted Newton with a smile. "You see, I've made good use of my time here in Cambridge, to learn more about today's astronomy and astrophysics."

"Isn't it strange," I continued, "that the galaxies don't move through space wildly and at huge speeds? We know today that they move through space in a civilized fashion, almost majestically. For instance, our own Milky Way and the galaxy closest to us—Andromeda—are moving toward each other but relatively slowly—at only a few kilometers a second."

"In my opinion, the regularity of cosmic motion as we observe it here can't be an accident," said Newton. "I imagine it has something to do with absolute space. It may even be identical with absolute space."

I wanted to avoid our getting lost in a fruitless discussion of principles on the question of absolute space: "Basically I agree with your last supposition. We observe in the cosmos that galaxies move in a regular fashion and don't follow some wild, chaotic trajectory across space. There is more order in this universe than the unbiased observer would have assumed. It is therefore reasonable to use a coordinate system that explains these motions simply. It wouldn't make much sense to use a system in which all galaxies were in orbit about one central point, and to define our rotating planet Earth as that point.

"For us here on Earth, all galaxies as well as the sun complete a rotation in twenty-four hours. This period, of course, has nothing to do with the galaxies but only with our modest planet. Astronomers have known this for quite some time. That is why they refer galactic motion to a coordinate system defined by our galaxy and having its origin at the center of our galaxy. As far as I'm concerned, we could establish this system as absolute space, the more so because with a system of that kind it's fairly easy to explain the motion of the other galaxies, since they move slowly with respect to our own galaxy.

"Let's start out by agreeing on this definition of absolute space, even though it looks a bit too phenomenological to you and certainly doesn't touch on the religious aspect. It will soon become apparent that questions of a completely new kind emerge in relativity theory; they'll force us to take another look at the notion of space."

I realized I was not being easy on Newton; I was calling essential aspects of his mechanical world view in question. But Newton was so eager to learn more about relativity theory that he decided not to put up an argument. Indeed, he himself suggested we should discuss the new theory right away.

Time was passing, however, and we agreed to set physics aside for a while and retire to a nearby inn for lunch.

Over the meal, we discussed many things that appeared to interest Newton. He was fascinated with my account of the first lunar landing, and wanted to find out about modern space research. We were in the best of moods when we left the inn. Neither of us felt like resuming our dialogue on physics. Instead, we enjoyed a stroll through Cambridge.

FOUR

A Dialogue on Light

Arriving at the park behind Trinity College, where I had taken a rest in the morning, we sat down on the lawn in a quiet place. My companion began to speak.

NEWTON

In one of the physics books I have consulted, it was mentioned that the issues in relativity theory are closely connected to the properties of light. So let's talk about light for a bit.

In my time, I was of the opinion that light consists of small particles that move through space at a great speed. A precise study of the motion of Jupiter's moons has made it possible to determine this speed, or, rather, the speed of those particles of light relative to absolute space—approximately 300,000 kilometers a second. I stress the word *approximately*. After all, we're dealing here with a rough estimate of a very great speed, which will depend on the speed at which the source of the light moves and on other circumstances, as the case may be.

HALLER

Wait a minute. Today we can measure the speed of light quite precisely. In one second, a light ray travels 299,792,458 meters. This speed is known to a precision of just about one meter per second.

NEWTON

Amazing! That kind of precision is hard to believe. We must find time for you to explain the ingenious methods that have permitted that measurement. But first of all, I'd like you to consider this: Assume I shine a flashlight in the direction of the tower you see over there. Light particles move from the lamp to the tower at a speed we'll call *c*, which is close to 300,000 kilometers a second. Now, I take the flashlight and run fast in the opposite direction, away from the tower.

The light particles leave the lamp with speed *c;* the lamp, however, is being displaced at a speed of 5 meters a second. We don't see the light particles at speed *c* but at speed *c* minus 5 meters a second.

This process is reversed when I run toward the tower, flashlight in hand. In this case, the light particles move, when seen by an observer at rest, at a speed greater than *c*. To be precise, their speed is *c* plus 5 meters a second. In my day, it would never have occurred to me that the speed of light could be measured to a precision of one meter per second. But a few minutes ago you mentioned that order of precision, which means that today we should be able to distinguish different speeds in light particles on the order of a few meters per second.

Which speed were you talking about just now—the speed of the light particles with reference to the flashlight or to a specific observer?

HALLER

I was simply talking about the speed of light as such. That's the speed denoted by the letter *c*. Whenever we measure the speed of light, we find the numerical value *c* as I gave it before. We'll never find another speed. The speed of light is a constant of nature.

NEWTON

That sounds absurd, inconceivable. You're a physicist, you should know that no speed can be a constant of nature. Any speed depends on the observer. The speed of light emitted by a moving ship or by a fast-moving car will be different from the speed of the light particles that issue from the flashlight I'm holding.

To maintain that all light particles move at the same speed is as absurd as saying all cannonballs fly at the same speed, regardless of the observer. That doesn't only contradict the laws I formulated in my *Principia;* it also runs counter to common sense.

HALLER

I agree with you. What I said is certainly contradictory to your laws; it appears to go against common sense as you invoke it. But it's true, nonetheless: light always moves at the same speed. This is not some notion of mine; it's a fact that's been proved experimentally.

NEWTON

When was the experiment performed? Who did it?

Newton, who had spoken with great excitement, couldn't wait

to find out the details. He sensed the danger that his whole edifice of mechanics would be unable to accommodate this experimental finding. I waited a few moments before continuing.

HALLER

Before we get to the experiments, let me say something about the nature of light. A while ago you spoke about light particles; they do exist, and today we call them photons.

NEWTON

What do you mean, "They do exist"? I talked about photon particles a long time ago in my *Principia*—of course they exist. But between you and me, I must admit I was not absolutely certain. There are some light phenomena, such as the diffraction of light at a narrow slit, that I wasn't able to explain with my hypothesis.

A few natural scientists on the Continent, especially in Holland and France, would have nothing to do with my theory. They insisted that light consists not of particles but of waves, which need some kind of medium in order to spread. This medium was thought to be a kind of ether filling the whole of space. The wave theory has some interesting features, apart from the notion of ether, which I have a real problem with. I might have accepted it if I had seen any possibility of making wave theory and my particle theory compatible—if some sort of compromise had been feasible. I kept looking for that. I finally did give up, and rightly so. After all, you yourself just told me that my theory of light particles turned out to be correct.

HALLER

Easy, easy, Sir Isaac. It's true that many experiments confirm the notion that light consists of minuscule particles—photons. But that doesn't mean your theory of light really was correct! The compromise you were looking for actually exists. It was discovered in 1905 by a young physicist who at that time was working at the patent office in my home town of Bern. He happens to be the same person who created the foundations of relativity theory.

NEWTON

My compliments to this man. What is his name?

HALLER

His name is Albert Einstein. We'll be mentioning him frequently. Einstein was probably the most important natural scientist of the twentieth century. He died in the United States in 1955.

NEWTON

I'm interested in this Einstein. You can tell me more about him another time, maybe tomorrow. For the moment, I'd like to know exactly how Einstein managed to find a compromise between waves and particles.

HALLER

I'm afraid I can't do that in detail right now, especially since those problems don't bear immediately on relativity theory. But let me clarify briefly: in the course of the eighteenth and nineteenth centuries, the wave theory of light slowly gained acceptance.

NEWTON

Aha, I suspected as much.

Newton looked at me full of anticipation. He was not disappointed but accepted my arguments calmly.

HALLER

Let's assume that light spreads in the form of waves, like a sound wave or a surface wave on a lake; we can then explain many light phenomena, such as diffraction, which we mentioned before. This assumption also permits detailed calculations of how light is guided through a complicated telescope. In this way we were able to build high-resolution telescopes, which have been particularly successful in astronomy. By the end of the nineteenth century, no one doubted that light is a wave phenomenon. But then some curious phenomena were observed, particularly in connection with atoms.

NEWTON

We don't need to speak about atoms now. In the last few days I've spent a fair amount of time on the theory of atoms. It's fantastic how scientists have managed to penetrate matter to the point of identifying atoms as the smallest components of chemical elements.

Are you surprised that I know quite a bit about atoms? I'm aware that atoms themselves consist of smaller particles, the nucleus and a shell made up of those tiny particles you call electrons. I've also spent some time on electrical forces, since they are what hold the components of the atom together.

There's still a lot I don't understand about atoms. Why, for instance, are some atomic nuclei so heavy? In my opinion, the

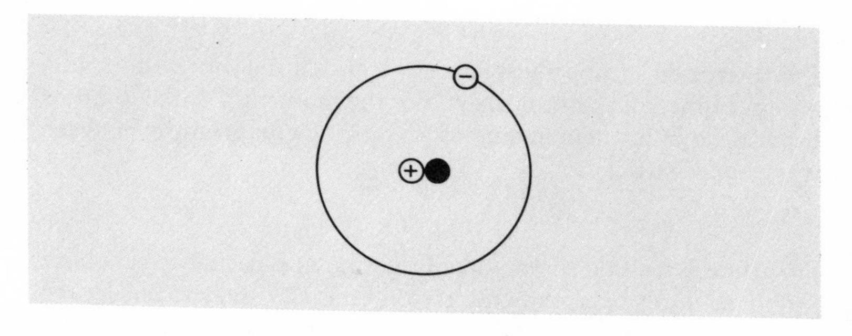

Fig. 4.1 Schematic view of a deuterium atom, consisting of a nucleus and an electron cloud. Its nucleus has a positively charged proton and an electrically neutral neutron (black circle); there is only one (negatively charged) electron in its cloud, orbiting around the nucleus. The stability of the atomic configuration is due to the electrostatic attraction between electron and nucleus.

More complicated atoms consist of several electrons orbiting around nuclei that contain several nucleons (protons and neutrons), where the number of protons generally equals that of the electrons in the shell or cloud, guaranteeing overall electric neutrality. The simplest atom is hydrogen, which has only one electron orbiting around one proton.

nuclei themselves should be made up of smaller particles. So the next question is, What forces hold these particles together? I don't believe they are electrical forces of attraction. Nuclei are too stable for that. I would think there must be some other force, a kind of nuclear force, quite distinct from electrical forces.

HALLER

You're perfectly correct to assume that nuclei are made up of even smaller particles. These components of the nucleus indeed exist, and are called nucleons. And nucleons are held together not by electrical forces but by far stronger ones, which are called strong nuclear forces, or are sometimes referred to as strong interaction.

Newton was fascinated. He was clearly delighted that I confirmed his points, one after the other.

HALLER

Let's get back to light, especially to the question of how to describe light phenomena by means of the particle-wave dualism. A while ago, a friend of mine published an article on that question in a popular magazine. I suggest you look at it. We might find the article right here in the college library.

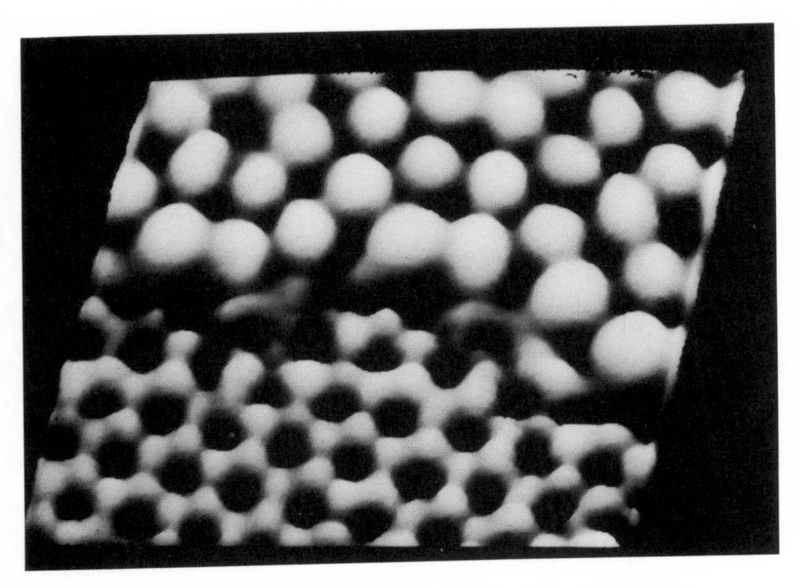

Fig. 4.2 Modern measurement methods allow us to visualize atoms. Shown here is an image created by a tunneling microscope. Atomic structure is clearly visible—here, silicon and silver atoms. What we actually see is not an image of the atoms themselves but rather the fields of electric forces due to the electron clouds as sensed by precision probes. These probes make use of a quantum effect called tunneling. (Photograph by R. Wilson, IBM Research Laboratory, San Jose, California.)

NEWTON

Good idea. You obviously think that I will understand it, even though three hundred years separate us from one another.

HALLER

That doesn't worry me. The article was written for people unfamiliar with physics. A nontechnical reader in the twentieth century will certainly have more trouble understanding it than the leading physicist of the seventeenth and eighteenth centuries.

NEWTON

Good. Let's find the article.

We were lucky; after a few minutes in the well-organized college library, I had a copy of the magazine in my hand. Newton started reading it right away. I planned to use the time for a short

Fig. 4.3 Albert Einstein at his desk at the Bern patent office. (Photograph, probably from the year 1905, courtesy of the Einstein Archives; by permission of the AIP Niels Bohr Library.)

walk through the town. We agreed to meet again two hours later by the fountain in the quadrangle.

[I don't want to risk my readers' getting lost at this point. In case they cannot locate the article I recommended to Newton, I am reprinting it here. I beg my readers' indulgence if it repeats some of the arguments that were covered in my discussion with Newton above.]

Newton's Reading: What Is Light?

It is the year 1904. A twenty-five-year-old civil servant, Albert Einstein, is working at the patent office in Bern, the capital of Switzerland. It is his job to review new patent applications.

Einstein's work keeps him busy. Still, he finds the time to ponder a number of physics problems that have occupied his mind since his student days in Zurich. Foremost among these is the nature of light.

For decades, there have been reports on unusual experimental effects of light. Nobody, including Einstein, has been able to make sense of them, but now he feels he is on the right track. He has an idea that will turn out to be the starting point of a fundamental change in our concept of the physical universe.

For quite some time, physicists had been working on the concept of light. And for good reason: next to the matter that surrounds us, light is the most obvious phenomenon in our daily lives.

But what is light? Is it a special kind of matter? One of the first physicists who tried to answer this question was the English physicist Sir Isaac Newton. In his main work, published at the end of the seventeenth century, Newton said that light consists of small particles. But that theory was not convincing. What, for instance, happens to light particles when light is being absorbed by an object? Are the light particles being "swallowed up" by matter?

Another view of light was proposed by a contemporary of Newton's, the Dutchman Christian Huygens. He believed that light, like sound, is a wave phenomenon, and that light waves propagate in a special medium that pervades all space; he called this medium ether.

Many properties of light can indeed be explained in this fashion, such as the refraction of light as it enters water or glass. This phenomenon is used in telescopes, for instance. It helped Huygens's idea to gain acceptance during the nineteenth century.

The wave theory of light triumphed when it became clear that light is simply a particular form of electromagnetic wave. This insight was gained at the end of the nineteenth century by the German physicist Heinrich Hertz. Electromagnetic waves, like the radio waves emitted by a transmitter, differ from visible light only by their wavelength. Thus, two important fields in the natural sciences were unified—the sciences of electricity and of optics.

The human eye registers only a small fraction of the electromagnetic waves—those with wavelengths of 0.38 to 0.78 of one-thousandth of a millimeter. The upper end of this range corresponds to red light, the shorter end to blue light. All other electromagnetic waves are invisible. This holds true for X-rays, which have wavelengths about one thousand times smaller than visible light. It is equally true of radio waves, which have much longer wavelengths, between one meter and several kilometers long.

Einstein's thoughts in 1905 originated in atomic physics. Matter consists of tiny building blocks—namely, atoms. They in turn are made up of even smaller particles—electrons and atomic nuclei. Electrons are the carriers of electric charges. The electric current going through a wire results from the motion of electrons in the wire; they hop from atom to atom.

Einstein was dissatisfied with the idea that, on the one hand, matter consists of atoms and thus has a granular structure, while light, as an electromagnetic wave, appears constant, lacking such granularity. However can we picture the interaction between atoms and light in this framework?

Heinrich Hertz, the discoverer of electromagnetic waves, was

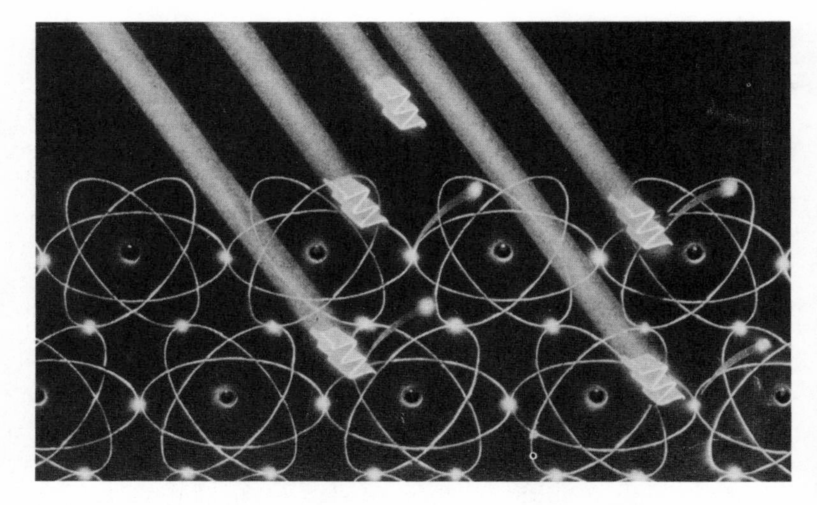

Fig. 4.4 The photoelectric effect: when photons hit a metal surface, electrons are ejected from it; applied voltage registers these electrons as a current. This is the operational principle of a photographic light meter.

the first to observe the odd effect that kindled Einstein's imagination decades later: when light falls on a metal plate, electrons can be ejected from the metal. This phenomenon is called the photoelectric effect. It later found many applications in technology—in the light meter of a camera, for example. Here, the light that falls on the camera, the "incident" light, releases electrons from a metal surface. The electrons, in turn, generate a measurable current. The stronger the incident light, the larger the indicated current: it tells us how to choose the exposure time.

Through incident light we can also "excite" atoms, so that they continue to radiate for a while. We see this, for instance, in the luminous numbers on certain clocks or watches.

Why are electrons released from metal? Like any electromagnetic wave, light waves contain energy, which is, in a manner of speaking, swallowed up by the metal. In nature, energy is never lost, not even in the photoelectric effect. Instead, the energy is transferred to the motion of the electrons that are ejected from the metal surface by means of the incident light

We might expect that the electrons will leave the metal quickly, with considerable kinetic energy, when the incident light is very intense; and that they will leave slowly when the incident light is

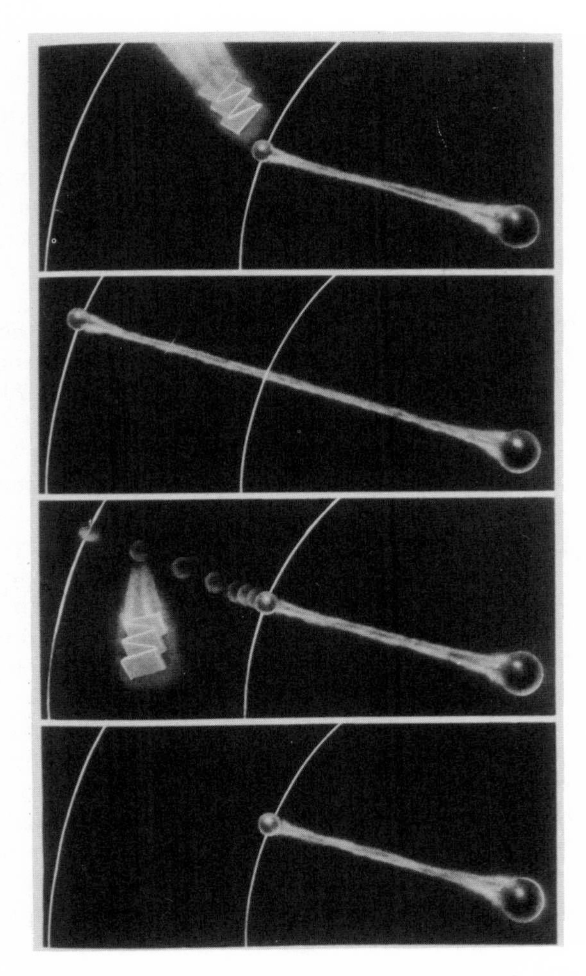

Fig. 4.5 Photons causing atoms to emit light. From top to bottom: (a) a photon hits an electron that is in an inner orbit around the atomic nucleus; (b) the electron absorbs the photon, and its increased energy permits it to pass into a higher orbit; (c) the electron does not remain in the higher orbit but drops back into the energetically favored orbit; (d) the energy gained by dropping back into a lower orbit is emitted in the form of a photon.

weak. After all, the stronger the light source, the greater the available energy.

But the observations of physicists yielded an unexpected result. The speed of the outgoing electrons did not increase when the incident light was made more intense, but the number of electrons did. If we increase the intensity of incident light by a factor of one hundred, the number of ejected electrons will increase one hundredfold. Yet there is no change in the speed or energy of these electrons.

There is, however, a way of changing the speed of the emitted electrons: we have to vary the wavelength of the incident light. If we use blue light, electrons are ejected with higher speeds than if we use red light, even if the intensity of the light is lessened. Physicists were puzzled.

Clearly, light striking a metal surface transfers its energy in a very specific fashion to the electrons that are ejected. Each electron obtains a given amount of energy. This amount depends not on the intensity of light but on its wavelength. It would appear, then, that a ray of light consists of many "atoms of light." What emerges is exactly the picture that Albert Einstein presented to an astonished community of scientists in 1905. It won him the Nobel Prize in physics in 1921.

Einstein's theory of light considers light a wave phenomenon, but one in which energy can be transported only in well-defined quantities. Einstein himself spoke of "light atoms," tiny "peas of light." These are what are now called photons, particles of light. Light, just like normal matter, ultimately consists of elementary constituent particles.

The question remains: Is light a wave phenomenon or a particle phenomenon? How should we depict a photon? Who is right: Huygens or Newton? Einstein's answer was that both of them were. The propagation of light is both a wave process and a particle process. We might more easily envisage it by breaking up a light wave into small sections, which correspond to individual photons—tiny wave and energy packets indefatigably hastening through space at the speed of light.

The energy of a photon depends only on the wavelength of the light in question. The shorter the wavelength, the greater the energy of the photon. The photons of red light are weaker in energy than the photons of blue light.

The energy of the "blue" photons is about three electron volts. (An electron volt, usually called eV, is the energy gained by an electron when it moves from the negative pole of a 1-volt battery to its positive pole. Because of the tiny mass of the electron, the energy is minuscule.) The energy of a "red" photon is only one-half that of a "blue" one: about 1.5 eV. Photons can be registered by the human eye only in an energy range between 1.5 and 3 eV.

The photons of radio waves have even less energy. A transmitter operating in the high frequency range has a wavelength of, say, 41 meters; this wavelength is more than 100 million times that of blue light. Therefore its photon energy is 100 million times smaller than the energy of the "blue" photons (about 3 eV).

X-ray photons have an energy about one thousand times that of visible light. That's why they are able to penetrate the human body and can consequently be used for medical purposes. But for the same reason, these photons can cause harm to human cell tissue.

The energy of photons can be of any magnitude. Photons with an energy of more than 10,000 eV are usually called gamma quanta. They are the building blocks of gamma radiation as it is observed in nuclear physics experiments. Intense gamma rays are also emitted by atomic reactors; they have to be shielded by walls made of lead or concrete. These gamma rays contribute to the devastating effect of atomic or hydrogen bombs; they are generated in the bomb's explosion in nuclear reactions.

Einstein's theory explains very simply why the energy of electrons in the so-called photoelectric effect does not increase with light intensity. When a photon strikes the surface of metal, its energy is absorbed by electrons in the metal. The affected electrons are accelerated and may be ejected from the metal surface. The speed of the ejected electrons is determined by the energy of the photon.

If we increase the light intensity, a greater number of photons strike the metal. As a result, the number of ejected electrons increases but not their speed. If we reduce the wavelength of the incident light by using blue light instead of red light, we increase the energy of the photons and, hence, of the ejected electrons. Thus, Einstein's theory makes sense of the originally puzzling photoelectric effect.

When Einstein suggested his hypothesis about light atoms—photons—his colleagues showed little enthusiasm. On the occasion of Einstein's election to the Prussian Academy of Sciences, Max Planck, who sponsored his membership, didn't conceal his critical attitude. He asked for indulgence for the fact that even as distinguished a physicist as Einstein might occasionally overshoot his aims with his speculations. Planck called the photon hypothesis an example of that excessive zeal.

Today there is not the slightest doubt about the existence of photons. They are just as much elementary particles as are electrons or the constituents of nuclei. Individual photons can easily be traced with modern detectors. It has been demonstrated that the retina of the human eye, which is basically nothing but a photon detector, can register individual photons of visible light. The retina is therefore sensitive to energies as small as a few electron volts.

Convincing proof of Einstein's hypothesis was furnished by the American physicist Arthur H. Compton. He took Einstein's idea very seriously, and set out to study reactions between photons and electrons. He argued that if photons are truly elementary particles, as electrons are, then it should be possible to observe collisions between photons and electrons, somewhat like the collisions between balls on a billiard table.

Let's assume that we are playing billiards, using small white balls and larger black ones. When one of the white balls hits a black ball that is at rest, the white ball will bounce off at somewhat reduced speed and at an angle. The black ball will also roll off, at a different angle. The speed and hence the energy of the white ball will decrease in the collision because part of the energy has been passed on to the black ball.

Compton, a practical man, compared the white balls with photons and the black balls with electrons. Electrons, the components of the atoms' shells, are in plentiful supply. To a good approximation, we can consider them at rest in atoms.

When we irradiate matter with photons of sufficiently high energy, some of the photons may hit individual electrons and will thus be diverted from their original paths. The electrons in question receive such a jolt that they are ejected from the material to which they were formerly attached and they can then be registered in a detector. In the collision with the electron, the photon loses some of its energy. So its wavelength changes.

This change of the photons' wavelength and the concurrent energy loss were demonstrated experimentally by Compton. Using X-ray photons, he irradiated normal matter such as a block of aluminum. In their collision with the electrons inside the aluminum, the X-ray photons were diverted from their paths and lost part of their energy—both phenomena being precisely measured by Compton. His results were a brilliant confirmation of Einstein's theory.

The collisions of photons and electrons detected by Compton furnished convincing proof that both photons and electrons are tiny particles. That raises the question of whether there is any difference between particles of light and particles of matter such as electrons or atomic nuclei.

An important difference can be found in the speeds with which these particles are seen to move. A common chunk of matter—say, a rock—may either be at rest or be moving at a given speed. This speed, however, cannot be unlimited. The laws of physics say that matter can never move at speeds greater than that of light.

In a vacuum, light moves at a speed very close to 300,000 kilometers a second. Photons move at the same speed, independent of their energy. A photon of visible light and an X-ray photon move at the same speed, although the X-ray photon has much greater energy. A photon produced by a nuclear reaction in the sun needs about eight minutes to travel from the sun to the Earth.

The speed of light is thus a universal quantity in nature. Here on Earth, in the space between sun and Earth, and in the vast reaches of space between the galaxies, photons always move at 300,000 kilometers a second. Constancy of speed is a special property of photons.

That is not the case for all other elementary particles, such as electrons. They behave just like larger chunks of matter. Like the rock mentioned above, the electron may be at rest or may be moving at a given speed through space, but its speed has to be smaller than the speed of light. In modern laboratories, elementary-particle physicists can easily accelerate electrons to speeds about 99.9% of the speed of light, with the help of complicated electrical and magnetic fields. (One such accelerator, the Stanford Linear Accelerator in California, is shown in fig. 4.6.)

Einstein, in 1905, was the first to recognize the universal significance of the speed of light. Both his theory of relativity and his photon theory were fashioned in this extraordinarily productive year.

But what, basically, is the difference between photons and electrons? It is their mass. The electron is an elementary particle with a tiny mass of 10^{-27} grams. Because of its mass an electron can exist at rest, just like a larger piece of matter.

A photon, on the other hand, is a particle with no mass. It

Fig. 4.6 Aerial photograph of the Stanford Linear Accelerator Center (SLAC), near Stanford University, California. The linear acceleration track, some two miles in length, starts at the foot of the Coastal Range and takes the accelerated particles straight to the SLAC experimental area. On the way, electromagnetic fields raise their speed close to that of light. (Courtesy of SLAC.)

does have energy, but it cannot exist at rest. Because it has no mass, a photon is compelled always to move at the speed of light.

The mass of an elementary particle is of great significance: it determines how fast a particle has to move in order to acquire a certain degree of energy. Yet physicists still don't know why some particles have mass, while others, like photons, have none.

In nature, photons play the role of energy carriers. The energy of the sun is radiated in the form of photons. Part of this energy is absorbed by the crust of the Earth, which transforms it into other forms of energy, such as heat. All this is possible only because photons interact with matter. They cannot simply penetrate matter; instead, they transfer energy to all electrically charged particles, as they do when they collide with electrons in the Compton effect.

It should be noted that photons can interact only with electrically charged particles. They have no interaction with particles

Fig. 4.7 American astrophysicists Arno Penzias (right) and Robert Wilson with their "photon detector," which resembles an old-fashioned ear trumpet. In 1965, these scientists proved that all space is filled with a homogeneous and isotropic electromagnetic background radiation. One cubic centimeter contains, on the average, some 400 photons. Photons are thus the most abundant elementary particles in the universe. (Courtesy of Bell Telephone Labs, Murray Hill, N.J.)

that carry no electric charge. There is an intimate connection between electricity and light, between electric charge and the photon.

Photons are not only very important particles in nature; they are also the most numerous particles in our universe.

In 1965, two astrophysicists, Arno Penzias and Robert Wilson, discovered a strange form of radiation that appears to be incident onto the Earth's surface in a uniform manner from all directions of space. These photons have a tiny energy of only 0.00002 eV. Today, we know that this radiation exists uniformly in all space, even between distant galaxies. The galaxies, the stars, and the planets are surrounded by an ocean of photons. There are, on the average, 400 photons per cubic centimeter. Our universe contains about 1 billion times more photons than electrons or nuclei.

Astrophysicists assume that this ocean of photons is a residue of the Big Bang—the explosion hypothesized to have occurred some 20 billion years ago, which not only generated all matter (the ashes of the Big Bang, as it were) but also filled the whole of space with radiation. Photons are therefore not only indispensable as transmitters and carriers of energy; they are the most abundant particles in the universe.

FIVE

Newton Meets Einstein

As we had agreed, I rejoined Newton two hours later in the quadrangle at Trinity. I was eager to find out how he had fared with the article I had recommended. He arrived shortly after I did. By his expression, I couldn't tell what his reaction was. I asked a little impatiently: "Did the article satisfy your curiosity about light?"

"On the contrary, my curiosity has increased. There are so many questions buzzing around in my head, I'm afraid it'll be quite some time before we get to relativity theory."

"I don't think so," I replied. "Most of those questions, particularly the ones that concern light, are directly or indirectly connected with the theory of relativity."

Newton answered: "Maybe so, but I can't judge that at this time. True, when I read that article, it became clear to me that not much remains of my original theory of light. The compromise suggested by Einstein between the wave concept and my own theory is an impressive achievement. When I wrote the *Principia,* I had an inkling of that sort of compromise, since it was clear to me that the wave theory of light had its merits. It offers a plausible explanation for interference and diffraction phenomena. Had I known more about electricity, and atomic and nuclear physics, and light considered as a form of electromagnetism . . ."

"My dear Newton, nobody will fault you for not publishing Einstein's hypothesis of light quanta in your *Principia.* Remember, Einstein did not reach his theory by pure reasoning. His success was possible because hundreds of physicists and engineers had amassed many experimental facts. Foremost among these were researchers as ingenious as your compatriot Michael Faraday, or the German physicist Heinrich Hertz. Besides, techniques for experimentation and observation had improved enormously in the meantime. Without these advances, the experiments concerning electrical and magnetic phenomena could not have been performed. In your time, those would have been impossible."

Newton relented a little. "All right, I admit that research techniques in my time left something to be desired. You are probably

Fig. 5.1 The parental home of the Einsteins in Munich, 12 Adelzreit Street, has survived war devastation. Einstein's father had his workshop in the backyard.

right; when I wrote the *Principia,* the time was not ripe for a complete theory of light. By the way, your friend's article made me curious about Einstein. Tell me more about him."

I proceeded to give Newton a short biographical sketch of the great physicist. My listener proved very interested, and never interrupted me.

Albert Einstein was born March 14, 1879, in Ulm, in southern Germany. A year later, his family moved to Munich. His father, Hermann Einstein, and his brother started a small business in the Bavarian capital, dealing in electrotechnical apparatus.

As a student at the Luitpold Gymnasium (high school) in Munich, Einstein showed a precocious interest in mathematics, the natural sciences, and philosophy. At age fifteen, he left Munich and moved to Italy with his parents. In 1894 his father transferred the company to Milan. The young Einstein went to Switzerland to gain the high school diploma necessary for admission to the university; he spent a year at school in Aarau before being accepted as a mathematics and physics student at the famous Federal Institute of Technology (ETH) in Zurich. He finished his studies in 1900 with a diploma qualifying him as "Instructor in Mathematics."

After applying unsuccessfully for the position of an academic research assistant at ETH, Einstein spent two years as a substitute teacher in various schools before joining the patent office in Bern in 1902, a civil service position. He remained in that position until 1909. Many years later, Einstein said that his time at the patent office was the happiest period of his life.

For him it was a time of important discoveries, introducing epochal changes in the natural sciences. In 1905, Einstein published several articles in the German periodical *Annalen der Physik* (Annals of Physics), which launched his international reputation. One of these articles, "A Heuristic View of the Generation and Transformation of Light," may have had a modest title, but it laid the groundwork for his theory of photons and led to Einstein's receiving the Nobel Prize in 1921. The other two articles, "Electrodynamics of Moving Objects" and "Does the Inertia of an Object Depend on Its Energy?" might be called the birth certificates of the special theory of relativity.

In 1909, in recognition of his achievements, Einstein was awarded a professorship in theoretical physics at the University of Zurich. In 1911 and 1912, he accepted professorships at the University of Prague and at his alma mater, ETH.

Einstein's university career was crowned by an offer to accept the prestigious position of a research professor at the recently founded Kaiser Wilhelm Society in Berlin. He held this post from 1914 until he emigrated to the United States.

After Hitler came to power in 1933, Einstein, who happened to be abroad at that time, decided to resign his position in Berlin and not to return to Germany. He never did return, and he died at Princeton, New Jersey, on April 18, 1955. He spent the last two decades of his life at the Institute for Advanced Study at Princeton.

Talking with Newton, I mentioned Einstein's major achievement during his Berlin years—the creation of the general theory of relativity, a theory of gravitation that goes far beyond Newton's concept.

Understandably, Newton was eager to learn more about it. But he concurred with my suggestion that it would be better to start with the basics of the original theory of relativity as published by Einstein in 1905, usually called the special theory of relativity.

After a short pause, Newton exclaimed: "I only wish I could talk to this Einstein in person!"

I was amused, and answered: "There is nothing I would like better. I never met Einstein personally. He died in 1955."

Newton smiled. "Don't forget, young man, that strictly speaking I'm no longer alive either. But I get your point. I would certainly like to see the place where Einstein worked in Bern. I would like to see the city he was living in when he developed that theory. It should be possible—after all, Bern is your home town, too. You may be surprised, but I mean this quite seriously: How about visiting Bern together? Better yet—how about flying there? Ever since I've been back in Cambridge and getting used to modern life I've wanted to take a plane trip, so why not to Bern? You said you had a few days free, so let's fly together."

Newton's idea made sense. I was tempted to join him on a trip to my home town.

"All right, we'll fly to Switzerland. I suggest we take a direct flight from London to Geneva, and then take the train to Bern. We could also visit the CERN particle research center in Geneva."

And that is what we did. That very evening we left for London. Newton stopped off at a hotel in Chelsea. I went back to tell my friends that I had suddenly changed my plans—that I wasn't flying to the USA but returning to Switzerland. They were surprised. Of course, I did not tell them of my meeting with Newton.

The next morning, I met Newton at the Swissair counter at Heathrow Airport. Overjoyed, he told me he had been there for over two hours observing the air traffic and the terminal. He was clearly looking forward like a child to the flight to Geneva.

Two hours later, around eleven o'clock, our plane started its descent to Geneva's Cointrin Airport. Newton, sitting by the window, was fascinated with the panoramic view of the Alps, dominated by the peak of Mont Blanc. I pointed out the narrow opening in the mountains below us, where the Rhone river breaks through the Jura chain. Soon we were flying over the southwestern suburbs of Geneva. Before we landed, I just had time to show Newton the CERN area at the foot of the Jura mountains.

Although the CERN research center isn't far from the airport, we decided to delay our CERN visit and take the train to Bern. It was early afternoon when we pulled into the main railway station of the Swiss capital.

The University of Bern is very close to the train station. The Institute of Physics, where I do my work, can be reached directly from the inside of the station by elevator. Newton and I took this elevator up to the large square in front of the university building,

Fig. 5.2 Aerial view of the European Laboratory for Particle Physics, CERN, in Geneva, Switzerland. The main building is at left next to the Super Proton Synchrotron ring, which accelerates protons and antiprotons. This ring is shown only by a circular line drawn into the photograph, since the accelerator is actually located in an underground tunnel. The broken line indicates the subterranean ring that houses the electron-positron accelerator LEP.

which offers a spectacular view of the city of Bern and the mountains of the Bernese Oberland. It was a fine, sunny day, and the snow-covered peaks of the Finsteraarhorn and the Jungfrau were sparkling. Newton gazed at this unique panorama with awe; he was pleased when I told him it had been "discovered" by countrymen of his more than a hundred years ago: that was the beginning of modern Alpine tourism.

We made a quick tour of the university quarter, walking past my laboratory. Since it was the semester break, few students were around. We climbed down some stairs leading to the Aarbergergasse, where Newton checked into a hotel close to the train station.

I needed to see a colleague at the institute, so I bade Newton good-bye. We agreed to get together for dinner at the Aarbergerhof, a restaurant popular with students and university faculty and staff. A few hours later, by the time we left the inn, an excellent meal and a fine Italian wine had put us into an exuberant mood.

Newton didn't want to wait another day to see the house where Einstein had lived. So I took him over to Bärenplatz—the square in the city center that is named for Bern's animal emblem, the bear. From there, we walked down the Marktgasse arcade (Market Street), past the municipal theater and the famous clock tower, and on to the street we were looking for, Kramgasse. We soon found ourselves in front of number 49, the house that Einstein had lived in shortly after the turn of the century. This was the birthplace of the theory of relativity.

One can hardly miss the entrance to Einstein's house, protected as it is, like so many others in Bern's downtown area, by a broad arcade. The column across from the entrance bears a plaque with the inscription, "In this house Albert Einstein created his fundamental treatises on relativity theory in the years 1903–1905." Einstein's apartment is on the second floor. In the seventies, the Einstein Society of Bern rented the apartment and transformed it into a memorial.

Newton wanted to see the rooms immediately. They are open to the public only during the day, but I had anticipated his wish and had made arrangements. A colleague of mine, an active member of the Einstein Society, had given me a key. It so happened the memorial was closed to the public for a few weeks, so I had had no trouble borrowing the key for a few days. Within minutes we were inside Einstein's apartment.

Nothing is left of Einstein's furnishings from his Bern period. There are just pictures, drawings, and documents connected with his life, along with the sparsest of furniture. Newton suggested I might want to take a walk around that part of town since there was nothing new for me to see here; he then turned to inspect the exhibits. I appreciated his wish to be alone at that time. I left the house and walked along the river Aare, past the Federal building, and on to the Old Town area.

It was about an hour before I was again walking up the narrow stairs to Einstein's apartment. To my surprise, Newton was not alone. He was engaged in an animated discussion in English, but his interlocutor's heavy accent gave him away as a German or, possibly, a Swiss-German.

When Newton saw me, he grinned and said: "You may recall that I told you in Cambridge how much I would enjoy being introduced to the theory of relativity by Einstein himself rather than by you. Well, it's a good thing we came to Bern. May I introduce you to our host, Mr. Einstein."

Perplexed, I took a close look at Newton's companion. True enough, the man in front of me was the living image of the person shown in a photograph on the opposite wall, standing in front of one of those old-fashioned upright desks in the patent office. This certainly was the real Albert Einstein, at the age of about thirty. He was of medium height, with broad shoulders, his heavy head framed by a mop of dark hair, and wearing a narrow mustache. His sparkling brown eyes were his dominant feature. The only difference between the man and the photo on the wall was the modern, somewhat worn gray suit he was now sporting.

We shook hands; Einstein, not without irony in his voice, said what a pleasure it was to encounter a Bern physics professor in this fashion. (He must have been alluding to the tensions that existed between him and the university faculty in his time.) Aside from that, Newton and Einstein acted as though it were the most natural thing in the world for the three of us to meet here in Einstein's apartment. Since I had just met Newton in Cambridge, I was not particularly surprised to run into Einstein in Bern. I had a feeling that Newton had anticipated this meeting before we left Cambridge—a good reason why he had insisted on the trip.

Einstein took it all in stride. "When I was in Bern," he said, "I founded an academy we called the Olympia Academy. Unlike other institutions of the kind, this one was useful—at least for its members. When I found out recently that there was a chance for me to return to Bern for a while, I had no idea I would find Newton in my apartment. How would you like to become a member of our academy, my dear Newton?"

"There's nothing I'd like better," replied Newton, "provided that you will tell me something about relativity theory. I assume the three of us will see a lot of each other over the next few days in this city. For that reason, I propose that we revive your old Olympia Academy. It will give an official stamp to our get-together."

I had no objections, of course, though I did feel a bit intimidated in the company of these two giants of physics. At this late

IN DIESEM HAVSE SCHVF
ALBERT EINSTEIN
IN DEN JAHREN 1903-1905
SEINE GRVNDLEGENDE
ABHANDLVNG VBER DIE
RELATIVITATSTHEORIE

Fig. 5.3 No. 3 Kramgasse (*Gasse* means a narrow street), the residence of Einstein and his family during his years at the patent office. It is here that Einstein worked out his fundamental ideas on the special theory of relativity and on the nature of light (see the memorial plaque, lower left). Einstein's old apartment has been maintained as a small memorial. (Courtesy of Albert Einstein Society, Bern.)

Fig. 5.4 The founding members of Bern's Olympia Academy. Left to right: Conrad Habicht, Maurice Solovine, Albert Einstein. (Courtesy Albert Einstein Society, Bern.)

hour, there was a revival of the Olympia Academy in Bern's Kramgasse. We celebrated it with a bottle of Montepulciano that Einstein managed to produce.

It had grown very late, so we decided to retire and to meet again next morning, here in Einstein's apartment. I took Newton back to his hotel. Einstein joined us for a while, then went his own way. It was obvious he was not going to spend the night in his apartment—there wasn't even a bed in there.

Next morning, Newton and I arrived at 49 Kramgasse at ten o'clock, to find Einstein waiting for us. We were ready for the first session of our academy. Einstein turned to me: "Both Newton and I are in a similar position. Both of us have suddenly been transplanted into this time slot. A lot of the everyday things I'm now seeing are incomprehensible to me, but I do have an advantage over Newton. Not nearly as much time has elapsed between my original stay in Bern and today as the corresponding time span for Newton. In the past few days I have been trying to understand what has happened in the meantime. Naturally I have concentrated on physics, and I have been helped by making good use of the library in the physics department, though I confess I have not made much progress yet. I'm telling you this right now, because I'm afraid there will be questions from Newton during our discussions that I will not be able to answer. I will rely on you to help me out."

Of course I agreed to do all I could. "First of all, Newton would like to know more about relativity theory. For this, your answers are the ones he'll need most. Why don't we ask Sir Isaac to get us started?"

SIX

The Speed of Light as a Constant of Nature

Newton began by describing the problems he had been having with a constant speed of light. Apparently, he hadn't stopped thinking about it since we left Cambridge. He now turned to Einstein.

NEWTON

How is it possible that light always travels at the same speed? In my mechanics it doesn't do that. The speed of an object depends, as you both know, on the observer. If the observer changes his speed, then the speed of the observed object must also change. The speed of any object is merely relative. It depends not only on the object, but also on the state of motion of the observer. Why should it be otherwise for light?

But that's not the only problem I have with light. We discussed your theory of light quanta before we left Cambridge. It appears to be a sort of synthesis between my own interpretation of light in terms of particles, on one hand, and an interpretation in terms of waves, on the other. I must congratulate you, Mr. Einstein, on that achievement. But if light is at least partially a wave phenomenon, I am wondering what medium it moves in. Ocean waves use the surface of the water as their medium; sound waves use the air. What then is the medium of light?

On top of that, I learned that light represents an electromagnetic phenomenon, and that there is no qualitative difference between a light ray, an X-ray, and a radio wave. All of them are electromagnetic waves, differing only in their wavelengths.

Now, if there is a medium in which electromagnetic waves can travel—say, an ether—the manner in which things move with respect to that ether is surely relevant. You can speak of a constant speed of light only when the traveling of light is measured in a reference system that is at rest with respect to the ether. I could accept a system of that kind. It would be compatible with my ideas of absolute space, which could then be determined by this electromagnetic ether. Hence my question: Does the ether exist?

Einstein listened attentively to Newton. He hesitated before answering.

EINSTEIN

I understand your objections, Sir Isaac. These same questions were on my mind for years. But there were others that kept me busy ever since I was sixteen years old. I didn't find the answers until 1905. What happens when the observer chases the light wave, following at the speed of light himself?

With an ocean wave, this can be done quite easily. Just the day before yesterday, in a movie, I saw someone engaging in a modern sport called surfing, standing on a board and "riding" a wave. He was moving at the same speed as the wave. Let's assume we could surf on a light wave. What would we observe in our reference system? At best we could observe the crest of our wave and the one next to it, and these would appear at rest. The picture we'd see would be static; it would be radically different from the one seen by an observer at rest who was watching light waves move by with the speed of light.

NEWTON

That does seem odd. As far as I know, electromagnetic phenomena can very well be described mathematically; as in my mechanics, the reference system plays no role. Qualitatively, light waves should look the same in all reference systems. It seems strange that in one case you get a dynamic picture, a traveling light wave moving, while in the other cases you get a static picture that doesn't change with time, like a painting on the wall. Something doesn't seem to make sense here. Maybe it's impossible to "surf" on a light wave, for some reason still to be determined.

Newton's argument made me smile, and Einstein winked at me. Both of us realized that Newton was on the right track.

EINSTEIN

What you just said is pretty close to the truth—or pretty close to my theory of relativity. We will shortly see that there can be no such thing as surfing on a light wave. But first let's get back to the problem of the ether, which, for a variety of reasons, kept a number of physicists busy in the late nineteenth century.

HALLER

You are probably speaking about the result of Michelson and Morley's experiment?

EINSTEIN

Yes, but not only that one. Still, let's talk about that experiment. It was based on a very simple idea. If there is such a thing as ether, you would expect the Earth to move through it while rotating around the sun. The Earth, after all, moves through space at a speed of about 30 kilometers a second. More precisely, we observe this speed in a reference system in which the sun is at rest.

Now, if the ether were at rest with respect to the sun, the relative speed of Earth with respect to the ether would be 30 kilometers a second. The observer on Earth would have an impressive gust of ether wind blowing in his face. Of course, he would be unaware of it, since we are assuming that there is no friction when the Earth moves through ether. Otherwise, the ether wind would have stopped the rotation of the Earth around the sun long ago.

We have assumed that the sun is at rest with respect to the ether. But even if that were not the case, we couldn't do without an ether wind, since the velocity—that is, the directed speed—at which the Earth moves changes constantly throughout the year. The direction of the Earth's motion at the beginning of summer is the opposite of what it is at the beginning of winter. So the ether wind can never be absent throughout the year; if it were, we would have to assume that it orbits with the Earth around the sun.

NEWTON

That makes it clear that we can measure the speed of the ether relative to the Earth. You merely have to measure the speed of light here on Earth in different directions and at different times of the year.

EINSTEIN

Congratulations! You just re-created the basic principle of Michelson and Morley's experiment. But let me say a word about the history of this important experiment: Albert Abraham Michelson was an American physicist who set himself the task of proving the existence of the ether wind while he was still in college. He

Fig. 6.1 American physicist Albert Abraham Michelson (1852–1931) (left) with Einstein (center) and Caltech president Robert A. Millikan (right). This picture was taken in the late 1920s, while Einstein was serving as a visiting professor at Caltech. (Courtesy of California Institute of Technology.)

did his first experiment during an internship at the Hermann von Helmholtz Institute in Berlin in 1881. It was a rather crude experiment, and his results were inconclusive.

Later, Michelson returned to the United States and, with his colleague Edward Williams Morley, a chemist, performed a much more sophisticated experiment. Its results were beyond doubt. This experiment was carried out from 1887 on. The two scientists built a piece of apparatus capable of measuring very small differences in the speed of light rays as they move in different directions.

NEWTON

Can you tell me what this apparatus was like?

Instead of answering, Einstein took a piece of paper and began to sketch (see fig. 6.2).

EINSTEIN

The idea of the experiment was to compare the speed of two light rays moving in different directions. Let's assume we generate this light from a given source, and let's arrange for this light to be

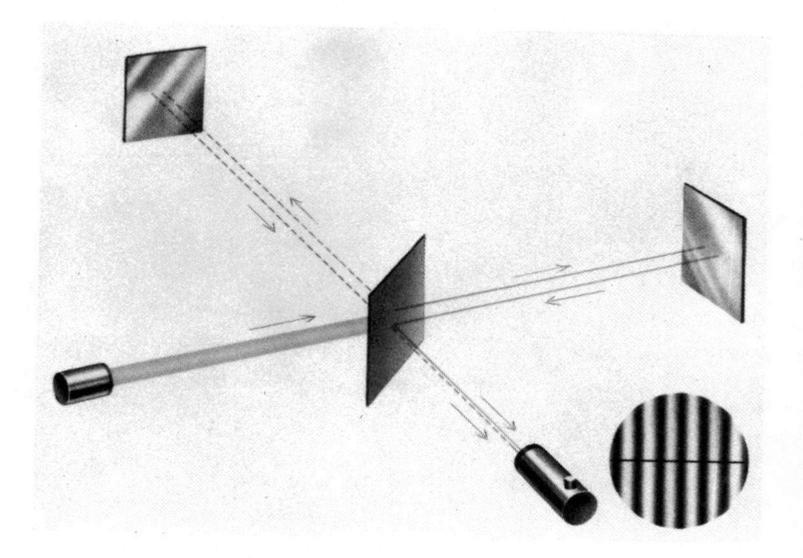

Fig. 6.2 Schematic view of Michelson and Morley's experiment. A monochromatic beam of light issues from the source at left and is then split by a silver-glazed mirror into two beams at right angles to each other. These two beams are reunited after reflection by two mirrors. The resulting photon beam is analyzed in an observation telescope. By making use of such interference phenomena as the structures shown in the circular inset, even the slightest differences could be detected between the travel speeds of the two light beams.

monochromatic, that is, of a given color. The ray travels, say, to the right and is split in half by a glass mirror. One-half of the light penetrates the glass and keeps traveling in the original direction. The other half is deflected by 90 degrees, and now moves in a direction perpendicular to its original motion. After traveling a few meters in their respective directions, both light rays will be reflected by means of two more mirrors. The returning light rays meet again at the silver-plated surface of the original glass mirror, and part of the light will be deflected sideways toward a telescope, through which it can be observed.

NEWTON

Let's see now: if the light particles of the two rays travel at different speeds, we'd detect slight differences in their arrival times at the final observation point. Part of the light would arrive earlier, another part a little later. But why do you need a special telescope for this observation? Wouldn't an accurate clock do the job?

EINSTEIN

In principle you're right. But there is a slight problem: Remember, we're dealing with a speed of about 300,000 kilometers a second, whereas the expected difference between the two parts would be on the order of only 30. The possible differences in arrival times would be minuscule, way below the level that can be measured with a clock.

So here we use a trick that exploits the wave character of light. When two light waves meet, they can either amplify or cancel each other, depending on which parts of the waves coincide—two minima, two maxima, or a minimum with a maximum. We use the term "interference phenomena." They can be observed with a small telescope, but I don't want to go into detail. Anyway, using this method we can observe and measure very small time differences.

NEWTON

You're tormenting me, Mr. Einstein! Come, tell me the truth! How great were the time differences measured by Michelson and Morley? Do they correspond to the orbital velocity of the Earth?

EINSTEIN (smiling mischievously and accentuating every word)

The result was zero. No difference was observed, although the experiment was so precise that they would have noticed the effect of ether even if the Earth moved through space at a speed of only 5 kilometers a second rather than the actual 30.

NEWTON

That means the speed of light is constant after all . . .

He had spoken very faintly, desperately attempting to hide his disappointment.

EINSTEIN

The speed of light is the same in every reference system, Sir Isaac—it's a constant of nature. Everywhere in space, here on Earth just as much as in a distant galaxy, light travels at the same rate.

NEWTON (turning pale)

My goodness! Light really is a crazy phenomenon! It seems impossible. How can light travel at the same speed in all reference systems? Think of that light-wave surfer of yours. Suppose somebody is actually riding along at the same speed as the light wave, doesn't that mean that light *can't* travel within his reference system at its original speed of 300,000 kilometers a second? What happens if the observer of a light wave moves at a speed greater than that of light? He would have to overtake the light wave, so that in his own reference system the light would hurry off in the opposite direction. But if I understand you correctly, that's impossible. It would be in direct contradiction to my laws of mechanics. I must admit, Mr. Einstein, this appears to be totally crazy, if not illogical. Please forgive me for sounding rude . . .

Einstein smiled and listened sympathetically. He hesitated to answer; so I took over.

HALLER

Professor Newton, you were just saying you feel light behaves in a crazy fashion. I would like to stress that not only light is afflicted with that kind of craziness but also the motion of normal bodies. We have discovered that your laws of mechanics are not exact—merely approximate. The faster an object moves, the greater the deviation. But the deviation is significant only when the speed of the object is comparable to the speed of light, *c,* which, as we know today, is 299,792,458 meters a second.

NEWTON

In other words, it isn't just light that behaves crazily, but all of nature?

EINSTEIN

My dear Newton, it's quite understandable that you should be disappointed when your laws of mechanics are put into question. But I beg you to keep a cool head. It has been proved experimentally that the speed of light is a universal constant of nature. There is no way around it.

NEWTON

That's just my problem. Please tell me how I can accept it. I'm prepared to make appropriate changes in my laws of mechanics if that's what it takes. That I can accept. But when you say there is a universal speed of light, I believe you are contradicting not only the laws of mechanics but, worse still, the basic laws of space and time.

HALLER

Not the basic laws of space and time but *your* laws of space and time.

NEWTON (sarcastically)

My dear sir, are you implying that one can only understand the crazy behavior of light by changing the structure of space and time?

EINSTEIN

That's exactly what he's implying, Sir Isaac, and that's also what I suggested in 1905. It's what has since been called relativity theory. You can understand light only when you redefine the concepts of space and time.

Hearing these words, Newton turned even paler. He looked totally amazed, an understandable reaction in view of the extraordinary impression Einstein's words must have made on him.

For a while we sat in silence. Einstein doodled absentmindedly, sketching abstract figures, relaxing in his easy chair. Finally, Newton took a quick look at his watch and said: "Gentlemen, I think I need a little time in order to digest what I have just heard. It's almost lunch time. Let's take a break. I would like to take a walk by the river, and then perhaps we can meet again for lunch at the Aarbergerhof."

We agreed, and soon afterward Newton left Einstein's apartment.

SEVEN

Events, World Lines, and a Paradox

During lunch there was no talk of physics; but you could see by Newton's face that he had thought up all kinds of questions regarding the constancy of the speed of light and its consequences. Einstein, on the other hand, was in high spirits; he entertained us with many stories from his Bern period. He kept mentioning the original Olympian Academy, whose members obviously lacked the dignity and solemnity usually attributed to members of an academy. He also talked about the many excursions they took together—to Lake Thun and other places. I then suggested we should take advantage of the good weather and arrange a similar outing for ourselves. Einstein agreed readily, and Newton had no objections.

After lunch we walked back to Einstein's apartment along Aarbergergasse and across Bärenplatz, and then started the afternoon session of our academy. I had great difficulty preventing Newton from bombarding Einstein with a long series of questions and problems. Einstein and I had agreed in the meantime that we should first familiarize Newton with a number of concepts and ideas that are needed for the understanding of relativity theory. Einstein assumed the role of lecturer.

EINSTEIN

Gentlemen, before speaking about relativity theory proper, I would like to explain a few concepts that play an important role in it. These concepts can also be discussed in the framework of classical mechanics—in your mechanics, Sir Isaac.

Let's look once more at space and time. As you know, space is three-dimensional in our universe. That's a fact we deduce from direct observation. Mathematically, it means that space can be described in terms of a three-dimensional coordinate system. Each point in space is described by three numbers, the three coordinates. We don't know why space is three-dimensional; at least I don't know of any compelling reason. Or has modern physics come up with one?

Einstein had addressed this question to me, and I felt compelled to answer.

HALLER

No, not yet. As far as we know, space could have more than three dimensions, without making the laws of physics meaningless. But I'm sure that a complete answer to your question will someday be found. The structure of our world and of space is probably determined by the structure of the fundamental laws of physics, but we don't know these laws completely. I am deliberately using the word "probably" because we're not sure about it. It could be that the three-dimensional character of space evolved more or less accidentally, shortly after the Big Bang. As a matter of fact, I've played with that notion myself, but I haven't come up with a concrete answer.

EINSTEIN (with visible amusement)

What do you think, Newton? I think we can safely say that physicists still don't know why space has three dimensions. It looks as if there is still some work for us to do. But let's be serious. We take it for a fact that space is three-dimensional. Time, on the other hand, is one-dimensional; it is uniquely defined by one number.

NEWTON (breaking in)

In the *Principia* I spoke quite intentionally of absolute space, and also of absolute time. I imagined I could place a clock at any point in space. I can also assume that all these clocks will show the same time at a given moment. The totality of these clocks will then describe one universal time, ticking along uniformly all over space; I would identify this with absolute time. The units in which we measure time are of course irrelevant—whether minutes, hours, or arbitrary fractions of these units.

EINSTEIN (interrupting)

Enough, enough, Newton! We all know your idea of absolute time. What I want to get at is some kind of unification of space and time. Let's look at time as a new and independent coordinate, roughly equivalent to the three coordinates of space.

NEWTON (breaking in once again)

Just a moment, Mr. Einstein, what exactly does that mean? You don't mean to assert that time is a part of space, do you? I must object most strenuously. Space is space and time is time. You can't mix them up. That would be like mixing up apples and oranges.

EINSTEIN

Careful, there! Nobody has talked of mixing up space and time—at least, not yet.

NEWTON (with slight irritation)

And I hope we'll keep it that way.

HALLER

Well, I'm afraid, Sir Isaac, that we will soon *have* to talk about that kind of mixing.

EINSTEIN (calmly, but firmly)

Easy, easy, my dear colleagues! I didn't suggest that the time coordinate should become a fourth space coordinate. We are simply introducing the time coordinate in addition to the three space coordinates. That doesn't make space four-dimensional. It merely defines a continuum of space and time, which we'll call space-time. It is a construct of three plus one dimensions, and just for once we can't simply add three plus one to make four.

I would like to give a simple example of this space-time continuum. Let's assume we are describing the motion of a spacecraft along the x axis of our spatial coordinate system. This description is not restrictive; we can always do it by translation or rotation, for we know that the spacecraft moves along a straight line in any case. The advantage here is that we can simply leave out the other two space coordinates, y and z. These are both zero along the trajectory of the spacecraft, and so are irrelevant for the description of its location at a given time.

NEWTON

You have simply made the problem one-dimensional.

EINSTEIN

Exactly. You'll see why this is useful. We can now describe the motion of a spacecraft by marking each point along the x axis, that is, along the trajectory, with the corresponding point in time at which the spacecraft moves through it. In other words, we establish a timetable, just like a railway timetable.

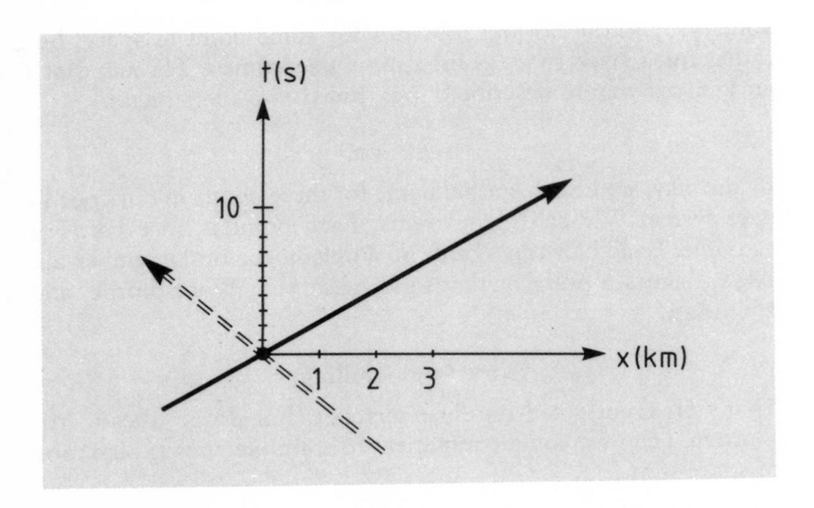

Fig. 7.1 Geometric representation of a two-dimensional space-time continuum, with one space axis x (in kilometers) and the time axis (in seconds). A spacecraft that is at point $x = 0$ at time $t = 0$ will follow the indicated bold straight line. The broken double line indicates an analogous motion in the opposite direction.

There is another method for introducing time as a specific coordinate. For this we need a rather unusual coordinate system, combining the x axis of space with a time axis.

Einstein began to draw a coordinate cross on a piece of paper.

EINSTEIN

I will now describe the motion of the spacecraft by giving each spatial point a corresponding time coordinate denoting the time at which a spacecraft passes point x. The result is a straight line in our two-dimensional—or, more precisely, our one-plus-one-dimensional—coordinate system. This way of describing the motion of a spacecraft has an obvious advantage: to find out where the spacecraft is at a given time, we no longer have to look up the times along the trajectory. We don't need the timetable of the spacecraft's motion; we just read off the x coordinate and the time coordinate for each point in the straight line of space-time.

NEWTON

Quite a trick. I must admit, your space-time coordinate system implies an amusing synthesis of space and time. Each point in your system now no longer stands for some point in space, but rather for a given space point x, at a given time t. For me, that's an unusual way to describe things, but it looks legitimate.

HALLER

By the way, we have a special name for these points in our coordinate system. We call them events. Each point is an event. For example, Isaac Newton's birth in Woolsthorpe on December 24, 1642, defines a point of that kind, with x = Woolsthorpe, and $t = 1642$.

NEWTON (smiling)

That's an event I can hardly remember. But do go ahead, Mr. Einstein. I can see you are impatient to continue your explanation of space-time.

EINSTEIN

The path of a spacecraft is a straight line in space-time. It represents a continuous sequence of events defined by the times at which the spacecraft passes each of the spatial points.

In a normal space coordinate system, the position of an object is defined simply by its location. In a space-time system, on the other hand, the object defines a line—the sequence of events through which the object passes. One such line or sequence is known as a world line. It contains all the information about the motion of an object in the past, in the present, and in the future.

NEWTON

I take it the name "line" has been chosen deliberately. You're implying that it needn't be a straight line?

EINSTEIN

Exactly. The world line of our spacecraft is a straight line because, according to Newton's law of inertia, it moves through space at a constant speed. This straight line has no beginning and no end. For simplicity, we'll assume that the motion of our spacecraft has no beginning and no end. That wouldn't hold, of course, for a real spacecraft, which must have been built some time or other.

I would like to mention another special case. Suppose a spacecraft is at rest at point x in a given coordinate system. In this case, the world line is once again a straight line, but this straight

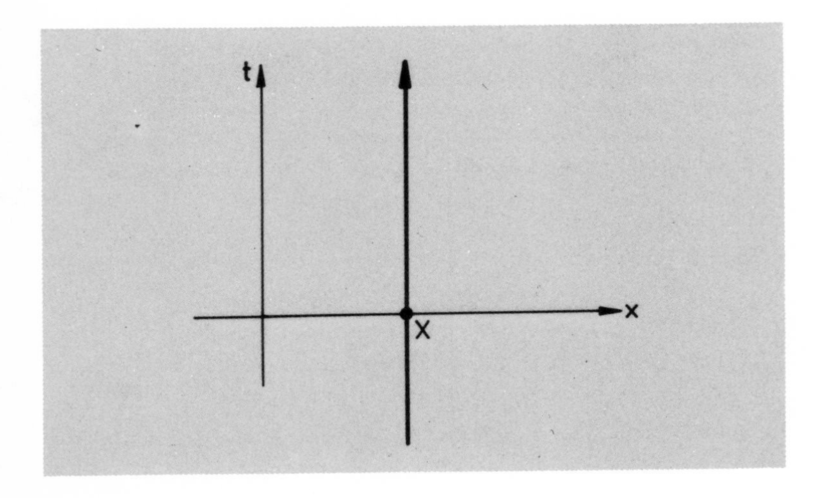

Fig. 7.2 The world line of an object that is at rest at point *x* appears as a straight line parallel to the time axis. Again, only one space axis is shown here; in three-dimensional space, the point will have three coordinates (usually called *x, y, z*).

line runs parallel to the time axis and crosses the space axis at point *x*.

The world line of an object moving through space nonuniformly and not on a straight line is clearly not a straight line. For example, the world line of a satellite moving in a circular orbit around the Earth is a spiral.

Many other curves can occur as world lines. They can hardly be described on a piece of paper, for their motion, like that of the satellite, may occur in three-dimensional space. To fully describe these world lines we would need a piece of paper on which four dimensions could be drawn, three for space and one for time, which is, of course, impossible. Even a three-dimensional model doesn't really help, because we have no way of representing time. For a mathematician, on the other hand, to describe that kind of world line in three-plus-one-dimensional space-time is no problem at all.

HALLER

I should mention that not all possible curves can be interpreted as world lines of physical objects in space-time. In a normal spatial coordinate system, every imaginable curve could be the trajectory

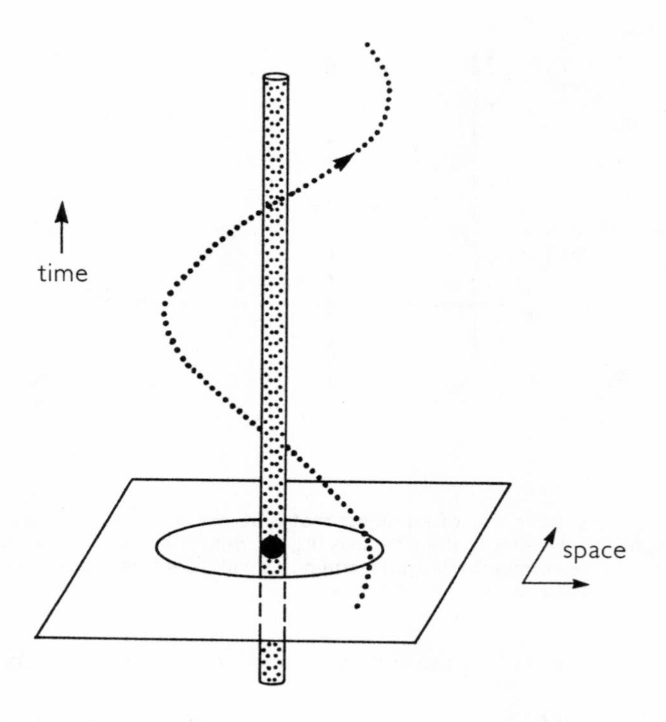

Fig. 7.3 World line of a satellite orbiting the Earth in a circular trajectory; it is a spiral winding around the straight world line of the Earth. Note that the Earth's motion around the sun is disregarded. Space is represented as a two-dimensional plane that shows a circular orbit for the satellite. The satellite's world line traverses this plane at one point of the circle.

of some object. In space-time, on the other hand, a material object can never follow a closed world line such as a circle.

Newton looked at me quizzically and pondered for a moment.

NEWTON

That makes sense. After all, a world line of that kind would mean that the object could pass through two different points in space at a given time. That, of course, is impossible for one and the same material object. I conclude that only those lines that, with time held constant, have one and only one set of space coordinates at any given time can be world lines of physical objects.

HALLER

That is one way to put it, and the mathematically correct one, so to speak.

EINSTEIN (taking over again)

I believe we have discussed space-time enough. Basically, it's nothing new for you, Newton, since we've been grounding our deliberations strictly on your mechanics up to this point. I'd like to leave that ground now and start talking about light.

Einstein again took the paper and drew a space-time coordinate system, with space denoted by the x axis alone (see fig. 7.4).

EINSTEIN

The origin of this space-time system, like any other point, represents an event—an event having both space and time coordinates at point zero. Now let's have something actually happen at this event point—say, we'll have a flashlight emit a light signal. This signal will travel at the speed of light, 300,000 kilometers a second.

Next, suppose we place an observer at a point x at a given distance from the origin, say, at $x = 300{,}000$ kilometers. It is his job to look for light signals. The world line of the observer, who is at rest, will be a straight line parallel to the time axis.

The moment the signal flashes, at time $t = 0$, the observer sees nothing. He sees it only when the emitted photons arrive at his point (x)—in this instance, after one second ($t = 1$).

Newton followed Einstein's words attentively, his head resting on his hand.

EINSTEIN

Next to the world line of the observer, we can draw the world line of the light signal. Since we are restricting ourselves for the time being to a one-dimensional space, the light signal will propagate along the x axis, in both the positive and the negative direction. There are, after all, only two directions, forward and backward. The photons that move along the positive axis reach the observer after one second. Their world line crosses that of the observer. The world line of the light signal is thus a straight line, and would be the same for any object with mass that was moving away from the origin at the speed of light.

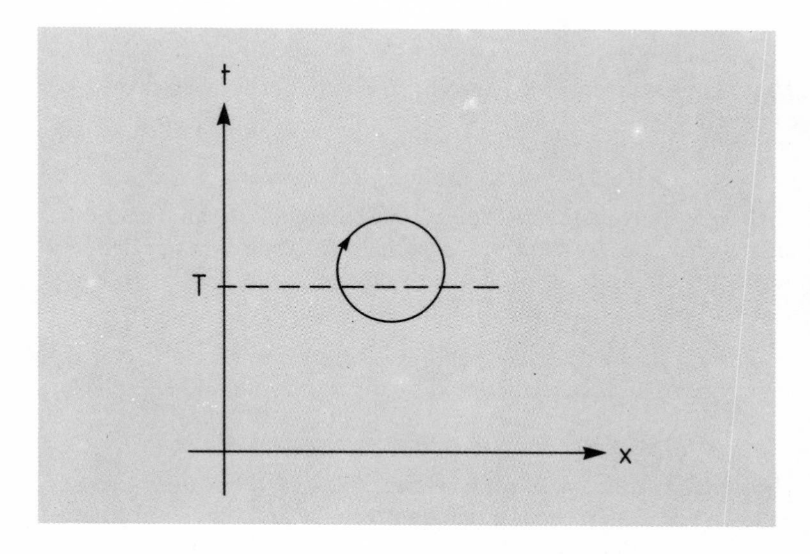

Fig. 7.4 A circle in the space-time continuum is one instance of a curve that cannot be the world line of a massive object. We see that at a given time—for example, *T*—the broken line, which contains all events that happen simultaneously, would traverse this world line twice. This cannot be, since an object cannot be in two places at the same time.

NEWTON (pondering Einstein's sketch)

If it is correct, as you said, that light always moves at a rate of about 300,000 kilometers a second, then the world lines of photons in space-time are always straight lines at a given angle with respect to the time axis. In your sketch, you chose the units in such a way that the angle is 45 degrees. That works only if your time unit is one second and your space unit isn't a meter, a mile, or a kilometer but the distance that light travels in one second—in other words, 300,000 kilometers.

HALLER

That makes sense. The distance light travels in one second is called a light second. It corresponds roughly to the distance from the moon to the Earth. This is a very small unit to an astronomer, who likes to think in terms of light years. A light year is the distance light travels in one year. Einstein chose a light second as the unit of length in his drawing.

EINSTEIN

You are right, Newton; the world lines of light in space-time are indeed remarkable. For the sake of simplicity, I am assuming that

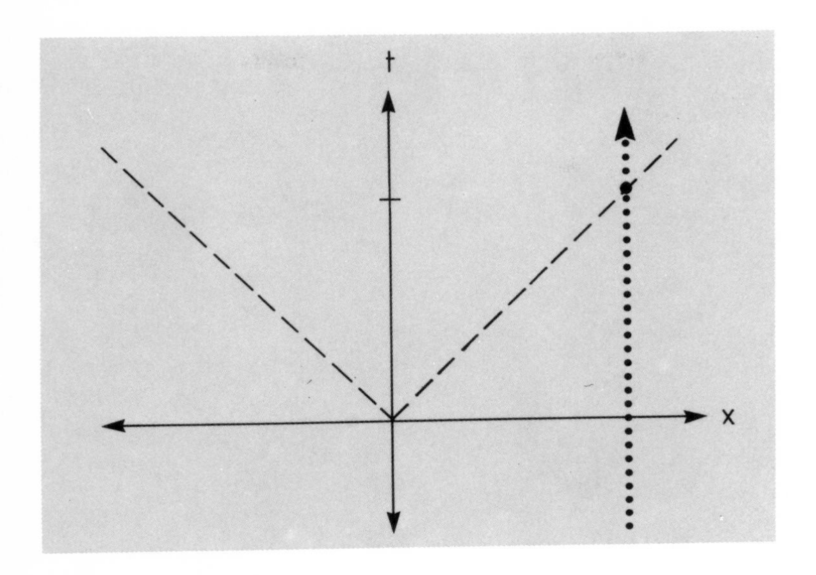

Fig. 7.5 A light signal emanates from event point $t = 0, x = 0$ and reaches an observer at distance 300,000 kilometers at $t = 1$ second (as marked on the time axis). At the event point, the world lines of the light signal (broken line) and of the observer (dotted line) cross.

my light signal starts from the origin of my space-time system. Its photons move along the two world lines that point to the right and to the left of the origin.

Now let's assume that we are not ignoring all other space dimensions. We'll add one more dimension, which can be described, for instance, by the y axis. With a little effort I can even illustrate that on paper.

Einstein now sketched a space-time with two space dimensions plus time, a two-plus-one-dimensional space-time (see fig. 7.6).

EINSTEIN

Again, I'm assuming that someone sends out a light signal at time $t = 0$, from location $x = y = 0$, that is, from the origin of space-time. The light can now move away in any direction in the two-dimensional x-y space, which is a plane. It is no longer restricted to two possible directions, but can move in infinitely many. So we can no longer say that emitted light follows a world line; rather, it has an entire plane of events. This plane assumes the shape of a cone resting on the origin of our coordinate system.

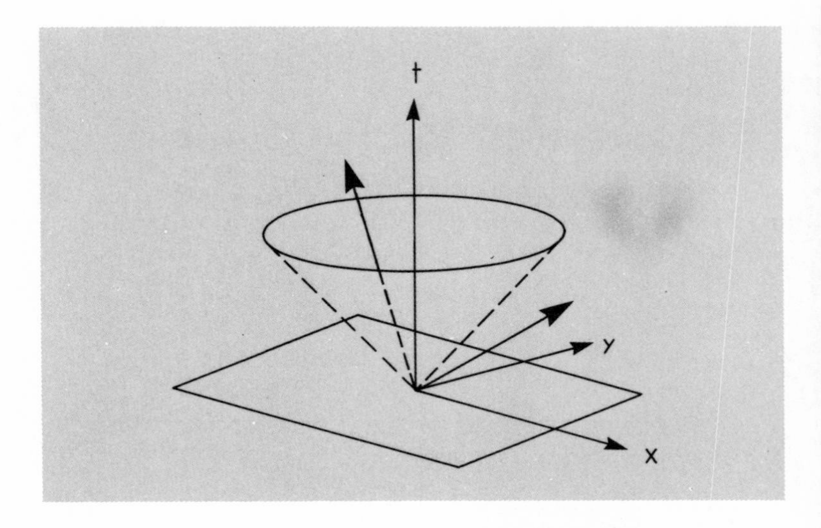

Fig. 7.6 The light cone in two-plus-one-dimensional space-time. The two arrows indicate two events, one inside the light cone, one outside.

NEWTON

I find it interesting that this light cone, if I may call it that, separates all of space-time into two parts. In one part are the events outside the cone, such as those in the *x*-*y* plane at time $t=0$. In the other part are the events inside the cone. These can be connected to the origin by world lines that correspond to the trajectories of objects moving at speeds smaller than the speed of light. To connect the origin with points outside the cone, we would need world lines of objects moving at speeds greater than the speed of light.

EINSTEIN

Congratulations, Newton! You are rapidly turning into a space-time expert. In no time you will also be an expert in relativity theory. You are quite right about the light cone, whose significance for the structure of space-time is due to the universal significance of the speed of light. Also, your dividing of space-time into two parts by means of the light cone is of the greatest importance in physics. But it's too early to go into that. Let's just point out that the light cone is an actual cone only in our example of a two-plus-one-dimensional space-time. In the reality of three-plus-one-dimensional space-time we have to deal with a generalized light cone that contains all events reachable by a light signal from the origin. Since light can move in all three space directions, this generalized light cone is a structure I can no longer sketch on paper.

NEWTON

I think you are giving me too much credit, Mr. Einstein. Up to this point we have discussed nothing that couldn't have been discussed in my *Principia*. The only new feature is the constancy of the speed of light—or, rather, its alleged constancy. I still can't believe that light moves at the same speed irrespective of the reference system. And I believe I can prove to the two of you that this can't be so. Last night I had an idea, and what we have been saying today about space-time has made it even clearer to me. So, gentlemen, I ask for your attention.

Einstein turned to me and smiled. He gave the impression that he knew exactly what Newton wanted to say.

EINSTEIN

We are very curious, Sir Isaac. What is your argument?

NEWTON

To be precise, I have two arguments. Here is the first. Let's assume that we are somewhere in space and observe the passing of a spacecraft. This spacecraft passes us at the speed of light, 300,000 kilometers a second. I agree that it is difficult to accelerate the spacecraft to that speed, but we can leave technical problems out of this discussion. It is only the principle we are interested in.

Now let a light signal be emitted from the spacecraft in the direction of its own motion. And here, gentlemen, comes my argument against the constancy of the speed of light. If it were constant in every system, it would have to be constant also in the rest system of the moving spacecraft. So the light signal moves away from the spacecraft with the allegedly constant speed of light.

But what will we perceive, observing all of this while at rest ourselves? Both the emitted light signal and the spacecraft are moving through space at the speed of light, that is, at the same speed along parallel trajectories. We have to conclude that the light signal can't move away from the spacecraft. But that contradicts what I said earlier in connection with the spacecraft. So there is a contradiction here; something must be wrong. I believe what is wrong is the assumption of the constant speed of light.

EINSTEIN (clearing his throat)

Sir Isaac, I agree that there is a contradiction in the situation you describe. But I'm not ready to follow your conclusion that the mistake is in the assumption of a constant speed of light. There is another solution.

NEWTON (frowning)

What solution would that be? Don't tell me the spacecraft is not allowed to move through space at the speed of light?

EINSTEIN (dumbfounded)

How could you know what I was going to say? Yes, indeed, it is impossible for the spacecraft to move at the speed of light.

I had the impression that Newton had anticipated Einstein's answer. My suspicion was soon confirmed.

NEWTON

I agree, my argument is valid only if it is possible, in principle, to accelerate a spacecraft—or some other material object that can emit a light signal—to the speed of light itself. If that is not possible, then there is no contradiction.

So, Mr. Einstein, if I understand you correctly, you are saying that it is impossible to accelerate a material object to the speed of light?

EINSTEIN

Correct. In the theory of relativity, or, you might say, in nature, the speed of light plays a fundamental role. Since it has been proved experimentally that the speed of light is the same in any reference system, there is only one solution to the paradox you have introduced: it is impossible in principle for a material object to move through space at a speed equal to or greater than the speed of light. It simply cannot happen. All objects move at lesser speeds than the speed of light.

Newton looked displeased. It was obvious he didn't like Einstein's answer.

NEWTON

How can you insist that no spacecraft can accelerate to the speed of light? I'll admit that it would be difficult technically, but it might be possible in principle. It needn't even be a spacecraft. Take a very small object, say, an atom or even a nucleus—it shouldn't be too difficult to accelerate it to a very high speed. You still contend that there is no way to accelerate such a particle uniformly until it reaches or even surpasses the speed of light? In my mechanics, there is certainly no problem with that.

HALLER (intervening)

Sir Isaac, nobody doubts that it can be done according to your laws. You can easily accelerate a car in ten seconds to a speed of 100 kilometers an hour. If you were to repeat this process every ten seconds, you would expect that the speed of light would indeed be reached after a considerable time.

But we now know that you can't keep repeating the process of acceleration. At high speeds—more precisely, at speeds on the order of the speed of light—it has now been demonstrated that your laws of mechanics are no longer valid. They must be replaced by the laws of relativity theory. According to the latter theory, it gets harder and harder to keep accelerating an object, such as the car we discussed. The closer its speed to the speed of light, the more energy is needed to raise its speed even by a small amount. And you can never reach the speed of light because it would take infinite energy to do so.

EINSTEIN

That's right. We can say the same thing in another way. A material object would need infinite energy to move at the speed of light. That's why no such object can exist; the energy of any given object is always limited.

HALLER

Let me give you an example, Sir Isaac. Shortly before we touched down at Geneva, I showed you the CERN compound. The huge CERN machine accelerates protons, the nuclei of the hydrogen atom. These particles have a positive electric charge, so they are being accelerated by strong electromagnetic fields while circulating inside a vacuum tube several kilometers long. If your laws of mechanics remained valid however rapidly the protons were moving, the CERN accelerator would push them beyond the speed of light. But this doesn't happen. The protons always move at a lesser speed than that of light, though their speed does come to within less than 1% of *c*.

Newton had jumped up and was pacing the room. There was a prolonged silence.

NEWTON

It's remarkable. I never thought the speed of light played such a fundamental role in nature. But how does the speed of light fit in here? The protons at CERN have nothing to do with light. What do they know about the speed of light that makes them unable to exceed it? The speed of light seems to mean more than just the rate at which light travels in space.

EINSTEIN

You are perfectly correct, Newton. Basically, the speed of light is a universal constant of nature in the true sense of the word. As we said before, it is of the greatest importance for the structure of space and time. You might call it the universal or basic speed. The fact that it is also the speed at which light travels has only secondary importance. The speed of light concerns everything, including the atoms our bodies are made of, which would appear to have little to do with light.

HALLER (turning to Newton)

It's quite likely that along with photons, the particles of light, there are other particles that always move at the speed of light—the neutrinos. These are electrically neutral and are related to electrons—we might call them the electrons' neutral brothers. They are produced in certain nuclear reactions.

NEWTON

Why did you say "likely"? Aren't you sure?

HALLER

No, we aren't. We haven't yet learned whether neutrinos have no mass, like photons, or whether they do have some mass. If they do, they could never move at the speed of light, any more than the CERN protons can. Be that as it may, the only thing I was trying to stress here is that "speed of light" is a somewhat equivocal term, as Einstein hinted earlier. One might choose to speak of the "speed of a neutrino," and again that would be a one-sided term. We're dealing, after all, with a fundamental constant of nature, and the fact that no speed can be greater than this constant has nothing to do with light or neutrinos. It's rooted in the very structure of space-time.

There was a moment of silence. Had we lost our thread? Then Einstein picked it up.

EINSTEIN

A while ago, Newton, you mentioned a second argument against the constant speed of light. How about that? Do you want to keep it from us?

NEWTON

Not at all, I was about to broach the subject. But I have to admit that what you said about the meaning of the speed of light won't go out of my mind. I'm not quite as certain now as I was an hour ago that my second idea is still sound. Still, here is my argument or, rather, my thought experiment. Suppose we are in outer space and observe three spacecraft passing us. This flotilla should move in a straight line, at an arbitrary, constant speed. Assume the speeds of the three spacecraft to be equal and to be smaller than the speed of light. Let the one in the middle be the command vessel with the main controls. The distances from the front spacecraft and the rear spacecraft to the command vessel are equal.

At a given time the command vessel sends out a signal. From the standpoint of an observer moving at the same speed as the spacecraft—a passenger in the command vessel, for example—we can describe it easily. The light signal leaves the command vessel and after a short while reaches both escort vessels. The signals arrive at the same time. Let me stress that: they arrive simultaneously.

Smiling, Einstein turned to me and winked. We both understood what Newton was driving at the moment he used the word "simultaneously."

NEWTON

Let's now look at the situation as external observers not moving along with the spacecraft. And this is the crucial point. You maintain that the speed of light is the same in any reference system, which means that both for the spacecraft and for our own refer-

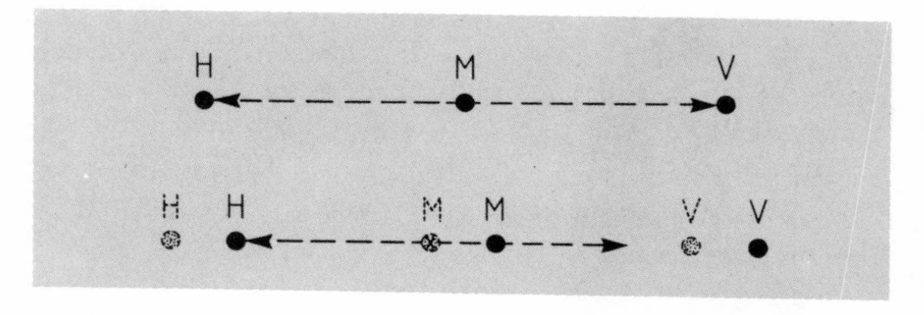

Fig. 7.7 Three spacecraft—M, V, and H—move uniformly along a straight line. The distance from M to V and from M to H is assumed to be the same. A light signal is emitted from M simultaneously in the directions of V and H. An observer traveling in M sees the signals arrive at V and H at the same time. Another observer, who is at rest, notes that the light signal arrives at H well before V notices anything. That is, the simultaneously emitted signals do not arrive at the same time at points V and H.

ence system the speed is the same: 300,000 kilometers a second. In a moment, you will admit that this is an absurd assertion.

Seen from our viewpoint, the light signal moves forward in the same direction as the spacecraft, and backward in the opposite direction, with the same constant universal speed.

Now, the light signal will take some time to reach the spacecraft. During that time, the rear vessel has been moving along the same path as the command vessel, so that the light doesn't need to move the entire distance between them to reach the rear vessel.

Now consider the light signal emitted in the forward direction. The front vessel moves on in the same direction while the light signal is traveling. So, the light signal has to move a longer distance than in the previous case.

Gentlemen, you surely have noticed how critical the situation has become: the signal reaches the rear vessel first, and then the front vessel. With that we have a totally absurd situation. Both signals arrive simultaneously in the reference system of the spacecraft but not in the system of the observer at rest.

But time flows evenly, irrespective of all reference systems. Two events that coincide in one reference system will also be simultaneous in every other. I conclude that there is something wrong about your constant speed of light, Mr. Einstein. The problem disappears as soon as you let the speed of light depend on the position of the observer. So I am of the opinion that there is something wrong about the Michelson-Morley experiment.

As he was speaking the last few words, Newton had jumped up and was stalking across the room, his gaze fixed on Einstein.

EINSTEIN

Take it easy, Newton! Come and sit down. Let's analyze this situation you call absurd, all three of us.

I should tell you that many years ago, when I was working out the basics of relativity theory during my time at the patent office, I too wrestled with those thoughts.

What you say is correct: the principle of a universal speed of light is incompatible with the assumption that two events that are simultaneous in one reference system—in our case the system of the spacecraft—happen at the same instant in every other reference system.

But I can't agree with your conclusion that the Michelson-Morley experiment was faulty. The results of that experiment and many similar ones are quite clear. The speed of light is indeed a universal constant, as we said earlier. It is a constant of nature. Once we are used to this postulate, the rest follows automatically. But I'll admit that it took several weeks, back in 1905, for me to accept the speed of light as a quantity of universal significance. We are therefore compelled to assume that two events that happen simultaneously in one system do not do so in another. In other words, if you move from one system to another, time changes.

In your specific example this means that there is one time in the system of the spacecraft and another time in the system of the observer at rest. To begin with, the two times are different. This is, of course, contrary to your mechanics. You would say that time elapses in a universal fashion, irrespective of the system. I, on the other hand, hold that there is no such thing as universal time, but there is a universal speed of light.

NEWTON

Are you sure, Mr. Einstein, that there is no other explanation? If what you say is true, it revolutionizes the structure of space and time. We agreed on that this morning, but I didn't take it quite so seriously then. It is only your answer to the question of simultaneity that has now made me waver.

HALLER

I can assure you, Sir Isaac, that there is no other explanation. We must abandon the idea of universal simultaneity of events. You realize what that means. We are about to develop a new concept of space and time, or rather of space-time. But let me calm your fears. Since the speed of light is very great, all deviations from your postulated structure of space and time will be very small, as long as the speeds in question are small in comparison with the speed of light.

Let's assume that our spacecraft moves at a rate of just a few kilometers per second—which, by the way, is about typical for a spacecraft. For the observer at rest as well as for the moving system, the two light signals will arrive simultaneously. The difference in the travel times of the signals is negligibly small.

Your ideas about space and time, which were so well expressed in your *Principia* and have held up well over two centuries, will not be completely invalidated. They remain perfectly applicable as long as the physical processes under consideration involve speeds much smaller than that of light. That includes virtually all the phenomena of present-day technologies. Only in phenomena involving speeds close to that of light will we notice sizable deviations from your space-time structure. We can already observe many phenomena of that kind—the motion of protons in the CERN tunnel, for example.

Newton had listened attentively. He rose, and declared he needed some time to think and to get some rest. So we closed our session; it was late in the afternoon, and we bade each other good-bye.

In so doing, Einstein said to Newton: "I fully understand why you are having a hard time giving up your space-time concept. I had the same trouble in 1905. Night after night I had to take long walks through the streets of Bern to work off my emotions. I couldn't sleep. Never would I have imagined in 1905 that, years later, the great Isaac Newton would have the same problem. I certainly wish you well for tonight. I'll see you tomorrow."

Newton thanked us and quickly left in the direction of his hotel. After sauntering for a while through the old town, Einstein and I sat down to have dinner in a little Italian restaurant close to Bärenplatz.

EIGHT

Light in Space and Time

Next morning I met Newton for breakfast at the hotel. His eyes were red and he looked tired. He must have spent much of the night trying to adapt his view of the world to the new information he had picked up in Einstein's apartment the previous evening.

After some small-talk I asked him whether he had changed his ideas on the speed of light.

NEWTON (with a forced smile)

I take that as a rhetorical question. Of course I have changed my view; what else could I have done after being bombarded with all those arguments Einstein and you let loose on me? I can't argue with facts, established by experiments. Physics, after all, is primarily an experimental science.

But you may rest assured that I have lost no time building the constant universal speed of light into my worldview! Maybe I should say I am still doing that. But there are remaining areas of vagueness which, I hope, we can clear up shortly.

I still have one problem with the universal speed of light. Originally, Michelson and Morley meant to measure the motion of the Earth with respect to the ether. They tried to do that by proving that the speed of light depends on the direction in which it travels. We discussed the result in some detail yesterday. No effect was found. And quite apart from this, there are several experimentally established facts in favor of a universal speed of light.

HALLER (interrupting)

Right, but let me add that we are speaking of light signals traveling in a vacuum. In other media, say in water or glass, light travels at a rate somewhat lower than in a vacuum. This is easy to understand. If light has to wend its way through the atoms of some medium like water, we would expect it to travel at a reduced speed.

NEWTON

Sure enough. But that's an effect of the specific material; it's not of basic importance. Of course, I was talking about light traveling in a vacuum. We have to conclude from the result of the Michelson-Morley experiment, negative as it was, that there is no such thing as ether. How can light travel in a vacuum with nothing to hold onto? Is it possible for light to travel without a medium?

HALLER

I was waiting for that question. But when you speak of light waves traveling in a vacuum with nothing to hold onto, that's not quite right. Light travels in space and in time.

Newton's eyes had lost all signs of weariness during this conversation. He now beamed.

NEWTON

Are you saying that space and time are really taking the role of the ether?

HALLER

To tell the truth, we still don't understand what light and matter, with their atoms and particles, really are. Some physicists suspect that light is really a hidden property of space-time—that it has a geometric meaning. In other words, space-time has not only three space dimensions and one time dimension but others too, which are not directly perceptible. They can be perceived only indirectly—through the phenomenon of light, for example.

Some physicists go so far as to say that space, time, and matter are merely different manifestations of some underlying geometric structure. According to that concept, the world is nothing but geometry.

Whether or not that interpretation is correct, today we see light as a kind of excited state of space or space-time, a particular property of space-time known as a field, or, more specifically, an electromagnetic field.

Newton's frown deepened. Clearly, something I said did not fit into his view of things.

NEWTON

I now understand what is meant by an electromagnetic field. Two particular cases, electric fields and magnetic fields, were studied intensively even in my time. The reciprocal attraction between

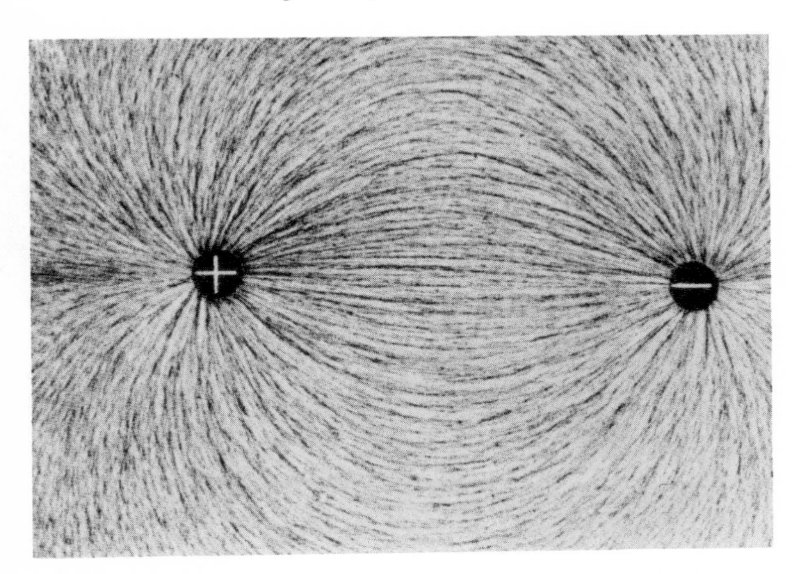

Fig. 8.1 Spheres with opposite electric charges attract each other; the origin of this force is to be found in the electric field that surrounds the spheres. The field lines can be made visible in various ways. Note that the lines always start and finish at the charged surfaces.

two objects that are electrically charged, one positively and one negatively, is simply a consequence of the electric field surrounding every charged object.

Magnetic forces work in a similar manner. The magnetic needle in a compass aligns itself in a north-south direction due to the influence of the Earth's magnetic field. When I was looking through physics books in Cambridge a few days ago, I also learned that we can have purely electric fields only when those fields don't change with time. As soon as anything changes—if, for instance, I move an electrically charged sphere back and forth—there will immediately be a magnetic field correlated with the electrical one. Parts of these fields may break off and leave the moving charge as electromagnetic waves.

HALLER

That's right. Radio waves are produced in almost the same way. A radio transmitter is essentially a device capable of producing time-dependent, pulsating electric fields. These oscillating electric fields then induce magnetic ones, and the combination of the two makes up the radio wave.

NEWTON

Since light waves are also electromagnetic waves, the question arises whether light can be produced in a manner similar to what you have just described.

HALLER

Absolutely. Let's take a look at a light bulb, for instance. The light here comes from a glowing wire that is being heated up by the electrical current passing through it.

NEWTON

But why does the wire emit light? Why does it glow?

HALLER

The electrical current that passes through the wire consists of electrons passing through the metal of the wire, or, better, being pushed through by the electrical tension or voltage. In the process they are constantly colliding with the atoms in the metal, that is, the electrons in the atomic shells, so that the metal heats up. In other words, all the particles are rapidly moving back and forth. The electrons in the metal are, of course, charged, and so they begin to emit electromagnetic waves, which in this case are light.

We continued to talk about various ways of producing light. It turned out that Newton had dismantled the lamp in his bathroom that very morning, just to find evidence that light really can originate in a long glass tube. It was, of course, a fluorescent bulb. I had to explain to Newton how it works, and that took us straight into atomic physics. Meanwhile, time had passed quickly, and we had to start toward Einstein's apartment to be there on time.

We walked along Kramgasse through Bern's inner city. As we walked, my companion raised some more problems he had thought about during the night.

NEWTON

If I understand you correctly, the electromagnetic field or, in the simplest case, the electric field that surrounds a charged sphere has an independent existence. It can be seen as a property of space or of space-time; it has a life of its own, which is only indirectly related to the charge. We can even make these fields visible by certain tricks, I found out from one of the books in Cambridge.

But what happens to the field if I suddenly remove the charge? It too would somehow have to vanish, for when there is no charge there is no field.

HALLER

You are quite right, the field will certainly vanish. You can make that happen experimentally in a laboratory. Just charge up a metallic sphere, then suddenly remove the charge. The field will disappear very quickly—it will happen at the speed of light, since all electromagnetic phenomena, not just light itself, travel at the speed of light.

NEWTON

Let's assume we are carrying out the experiment you have just described. You charge up your sphere, and I'll measure the electric field; or, rather, I'll measure a force of attraction at a given distance, maybe at 10 meters. Now you remove the charge. At first, if I'm not mistaken, I won't see any difference; the field needs some time to disappear. But light needs only 33 billionths of a second to travel 10 meters. It will take that fraction of a second for me to see that you removed the charge.

HALLER

You're right in principle. But it's virtually impossible to do the experiment because I wouldn't be able to remove the charge in such a very short time.

NEWTON

Of course, but I'm interested only in the principle. Now let's get to my real problem, which has nothing to do with electromagnetic phenomena but instead has to do with gravitation. You know that I formulated the law of universal mass attraction in my *Principia.* Two objects with mass attract each other; the strength of that attractive force depends both on the mass of each object and on the distance between them. The larger the mass, the stronger the force; the greater the distance, the weaker the force.

It was my original belief that this implies a long-range action of one body on another. In other words: the sun attracts the Earth because the gravitational force of the sun acts directly upon the

Earth across the distance between the sun and Earth. But I admit that I was not comfortable with the notion of this action over a long distance. And now that I know a lot more about the forces of electricity and magnetism, I have become still more uncertain. Can gravitational force really act directly across distances much greater than that between Earth and the sun?

HALLER

Your doubts are legitimate. According to our findings today, a very long-range action is not possible. The force of gravitation, like electrical force, is now understood as mediated by a field of force—in this case a gravitational field—surrounding every object that has mass. The mass of the object, in a manner of speaking, influences the space that surrounds it.

NEWTON

Does that mean that there are gravitational waves, like electromagnetic waves?

HALLER

They should exist, but to this day we have not found unequivocal proof. The following thought experiment suggests that they do. Let's assume we suddenly remove the sun. It wouldn't be easy, of course, but at least we can hypothesize. After all, there are periodically gigantic stellar explosions in space, and when they occur, masses about the same size as that of the sun are hurled over vast distances. Our hypothetical sudden removal of the sun can be compared with explosions of that kind.

NEWTON

I take it you are referring to supernova explosions like the one sighted in February 1987 in the Large Magellanic Cloud?

HALLER

You are well informed about astronomical events of modern times, I see. Yes, I am referring to supernova explosions. What do you think an observer on Earth would see if the sun suddenly disappeared?

NEWTON

For an Earthbound observer, the sun is especially important for two reasons: first, it gives us light and thermal energy; and second,

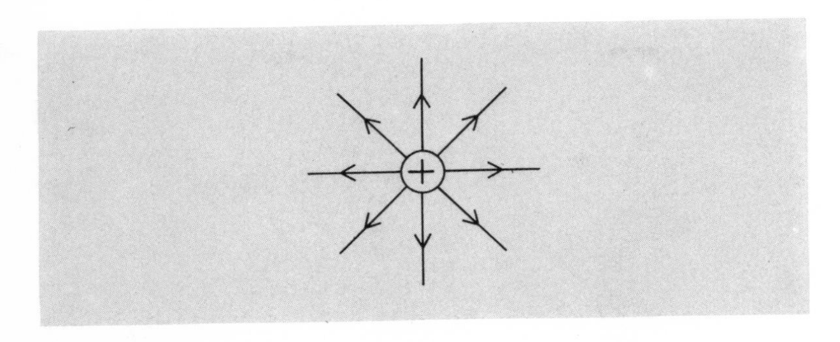

Fig. 8.2 The electric field around a charged sphere is suddenly removed. As a result, an electric shock wave radiates from the sphere at the speed of light. The electric field slowly decreases and finally disappears.

the sun's gravitation forces the Earth to move in an almost circular orbit around it. Without the sun it would be totally dark on Earth, and our planet would fly off into space.

Light needs about eight minutes to travel from the sun to Earth. That means an observer on Earth could enjoy the sunlight for eight minutes after the sun had been removed.

HALLER

Yes, just for eight minutes. But what about the Earth's orbit around the sun?

NEWTON

That's a tricky question. Before our discussion I would have said the Earth would cease its orbit immediately after the sun's disap-

Fig. 8.3 On February 23, 1987, astronomers witnessed a very rare event: a supernova exploded in the Large Magellanic Cloud (one of the small companion galaxies of our galactic system), visible only in the southern hemisphere. The bright supernova can be seen toward the top right in the picture.

A supernova explosion is a rapid collapse of a massive star coupled with the emission of a vast amount of energy. Within fractions of a second, electromagnetic shock waves are emitted. Given that light takes 160,000 years to travel from the supernova site to Earth, the explosion seen in 1987 happened some 160,000 years ago, in Paleolithic times. Since it must have shaken up the web of space-time, the resulting gravitational waves would have been observed on Earth if there had been a sufficiently sensitive detector in operation. Should a similar event happen in the foreseeable future, we would be better prepared: new detectors now being developed using laser techniques have the required sensitivity. (Courtesy of ESO Munich and La Silla, Chile.)

pearance and would take off on a straight path. But if there is no long-range action—and I'm now convinced there isn't—that obviously won't happen.

HALLER

You're right, it's impossible. True, we can make the sun disappear—at least in our thoughts—but that doesn't make its gravitational field disappear instantaneously.

NEWTON

In other words, what we're talking about is very like the former example of an electrically charged sphere that suddenly disappears.

HALLER

It's totally analogous.

NEWTON

Then I understand. The gravitational field, like the electrical field, is removed at the speed of light.

HALLER

Right again. The removal of the field happens in the form of a shock wave similar to the wave we generate when we throw a stone into a pond. But now we're talking of a gravitational wave that moves out in all directions like an ever enlarging sphere, at the speed of light, with the former location of the sun as its origin. It will take eight minutes for the wave to reach the Earth, and when it does, the Earth will start to move in a straight line. After a few hours, the wave will reach the outer part of the solar system. A few years later it will reach the closest stars, and after thirty thousand years it will reach the center of the galaxy.

The star that turned supernova in February 1987 must have emitted powerful gravitational waves, which reached the Earth at about the same time as the light signals from that explosion, that is, on February 23, 1987. Unfortunately, nobody could register them because there were no appropriate detectors in operation. If we should have another supernova explosion in our galaxy sometime in the next few years, we would be better prepared.

We walked in silence for a few minutes, passing the famous belltower. Newton was lost in thought. Suddenly, I heard him murmur, "So it's true after all—electricity, magnetism, and gravitation, all tied to fields, to properties of space-time. Simple, but ingenious. I did suspect it, that time in Woolsthorpe. Had I only known of the overriding importance of the speed of light!" Newton murmured on, but I could no longer understand what he was saying. He must have been talking to himself. So I didn't ask questions. By now we had reached Einstein's house.

The bell struck ten from the tower. We were half an hour late. Einstein received us at the door without mentioning our tardiness.

NINE

Time Dilation

While waiting for us, Einstein had prepared tea, which we enjoyed as soon as we arrived. The creator of the theory of relativity then turned toward me.

EINSTEIN

Mr. Haller, you recently mentioned the impressive precision that has been reached in the measurement of the speed of light. I noted down the number you mentioned: light travels at 299,792,458 meters per second. Now, given that this speed is a universal constant of nature, couldn't we use it to establish a connection between units of length and units of time, between meter and second? In other words, we could dispense with the kind of measurement based on an arbitrary standard, like the standard meter that is being preserved in Paris; instead, the path traveled by light in a given time could serve as the unit. Astronomers have been doing that for some time. They state distances not in kilometers but in light seconds, light minutes, and light years. That could work only if time were measured with great precision, of course. So my question is: How exactly can we measure time?

HALLER

Your suggestion makes a lot of sense. It has in fact been used for years. There is an international agreement that defines a meter to be the distance traveled by light in precisely one 299,792,458th of a second. Logically, this standard also defines the speed of light. So it doesn't make sense to measure the speed of light to a precision greater than that of 1 meter per second—let's say, down to centimeters or millimeters per second. The standard has now been fixed once and for all.

A perfectly accurate measurement of the speed of light won't give us additional information. What it will do is pin down the meter as the basic unit of length. Of course, according to the definition based on the distance traveled by light in a given fraction of a second, the newly defined unit of length was identical to the meter as it was known at the time. The main point of this procedure is that we now have a standard that can easily be reproduced anywhere. It is independent of numerous sources of error, which are no doubt unavoidable as long as we base the definition of a meter on some metal rod, however well preserved. Just think of the tiny notches we use to mark precise lengths on a rod of that kind. Under the microscope they are no longer hairlines but look like mountain valleys. We can avoid these uncertainties by indirectly measuring the unit of length by way of the constancy of the speed of light.

You can see that the new definition of the meter as the unit of length is a direct application of the universality of the speed of light and, hence, of the theory of relativity. Now let's turn to the measurement of time. In your lifetime, Sir Isaac, time was defined by astronomical standards. A second was determined as a certain fraction of a year, that is, of the time the Earth needs for one orbit around the sun. That's clearly inadequate for very precise measurement of time. A famous countryman of yours, James Clerk Maxwell, indicated more than a hundred years ago in his *Treatise on Electricity and Magnetism* that we should use the extensions and oscillations of atoms to establish precise units of space and time.

NEWTON (interrupting)

That sounds very sensible. Atoms have the same structure throughout the universe, so there would be no difficulty in reproducing the units of length and time anywhere.

HALLER

It's become apparent that atomic physics can help us to measure time very precisely—much more accurately than distance. In the 1930s the vibrations of quartz crystals were used to measure time. They furnish the standard for the quartz clocks and watches that are so popular today.

Even more accurate are what are called atomic clocks, in which the oscillating pendulum has been replaced by a beam of oscillating atoms. Today, we generally use the atoms of the element cesium for these clocks. The oscillations of cesium's atoms are the same everywhere, and so the duration of a second is fixed by a certain number of oscillations—quite a large number, of course.

[For the information of our readers: the number of oscillations per second is exactly 9,192,631,770. It is therefore easy to calcu-

Fig. 9.1 The cesium atomic clocks CS-1 (left) and CS-2 (right), located in the Atomic Clock Hall of the German Federal Institute for Physical Technology, Braunschweig. These clocks constitute the primary time reference for Germany. The frequency of an electromagnetic resonator is stabilized at the desired value by means of a beam of cesium atoms. This can be done because the atoms of a given element display identical oscillatory properties. The resonator can be tuned by an appropriate coupling of the resonator to the atoms. (Courtesy of PTB Braunschweig.)

late that light travels a distance of 3.26 centimeters in the time a cesium atom takes for one oscillation—see fig. 9.1.]

EINSTEIN

What precision has been reached in the measurement of time?

HALLER

There is a relative error of about 10^{-14}, even in the best laboratories doing work in this field. In Germany, for instance, that is the lowest level of error that has been achieved by the Federal Institute of Technical Physics in Braunschweig. The error implies that an atomic clock would be off by at most one second in a time span of 10^{14} seconds. That's only one second in about three million years! So we can justly say that time is the physical quantity we can measure with the greatest precision. Now, by enlisting the constant speed of light, we can translate this precision to the measurement of distances.

NEWTON (impressed)

That's amazingly precise! And it works only because of the universality of the speed of light. So we can safely assume that calibrating units of length amounts to the same thing as measuring the travel times of appropriate light signals. You see, my dear Einstein, your principle of the constant speed of light is being put to good practical use.

HALLER (interposing)

Today's discovery is tomorrow's calibration. This has become proverbial among physicists today. And it's what moves the sciences forward; yesterday's discoveries are taken for granted as the preconditions for new insights.

NEWTON (returning to our discussions of the day before)

Our conclusion last night was that the idea of the universal speed of light makes sense only if we reconsider the basic concepts of space and time. We have already seen that the answer to the question whether two events are simultaneous or not depends on the coordinate system used by the observer. This is in direct contradiction to the ideas of space and time that I formulated in my *Principia*.

Last night, I grappled with this baffling consequence. I came to the conclusion that we have to introduce a particular description of space and time for every system of coordinates, for every inertial frame of reference. A person in a moving train, for example, will then have his own private definition of time and space; it will be connected with, but not identical to, the description of time and space used by an observer at rest.

That doesn't imply that the train and the observer exist in totally different worlds. They share the same world, the same space-time. It's just that there are various ways of describing space and time. In other words: space and time are relative, the different ways they are described depend on the state of motion. I assume that is why you named your theory "the theory of *relativity*."

EINSTEIN (looking appreciatively at Newton)

You must have given all this a great deal of thought last night, Newton. I can't find any fault with your conclusions. Many years ago when I was working out my theory, I tackled the problem in a similar way, but much more slowly. In one night you solved something that took me days, if not weeks.

As far as the name of the theory is concerned, you're not quite correct. Calling it "the theory of relativity" wasn't my idea. It was suggested by others. I didn't like the name at first—I thought it sounded too complicated for a set of facts that are really quite simple. More to the point, the name isn't really descriptive, since my theory is based on the universality of the speed of light. In classical mechanics the speed of light is relative; it depends on the motion of the observer. But in my theory it is absolute. So my theory might have been called the theory of the absolute. You see—everything is relative, even the naming of theories.

NEWTON

If describing space and time really does depend on the observer's state of motion, one should be able to do it in terms of the observer's speed. I tried to work this out last night, but I didn't get very far. So perhaps we can discuss it now. Mr. Einstein, could you give us a short talk on your ideas of relativistic space-time?

EINSTEIN

My dear Newton, I consider it an honor to present to the founder of classical physics, the creator of classical mechanics, the further development of his own ideas, for that is what relativity theory amounts to. First, let me briefly repeat my principle of relativity, on which the theory is built. For two observers moving uniformly and on straight lines relative to each other, the same laws of physics apply. In particular, the speed of light is identical for both observers.

HALLER

Sir Isaac, you can see that Mr. Einstein's principle is a direct continuation of your ideas. In your mechanics, the same mechanical laws apply to two observers moving uniformly toward each other—this is what we now call Newton's principle of relativity. It makes no difference whether we are doing the research in a laboratory at rest, in a moving train, or in a fast-moving airplane. What's new about Einstein's theory is his assertion that his principle of relativity applies not just to mechanics but to all of physics. It includes electrodynamic phenomena and therefore all processes that involve light. And that implies that the speed of light is a universal quantity.

NEWTON (ironically)

I can certainly accept the principle you have just called a continuation of Newton's principle of relativity. If we go on like this, relativity theory will turn out to be an idea that for all intents and purposes was included in my *Principia.*

EINSTEIN

You came very close to it. Had you been told at the time that the speed of light is identical in every reference system, you probably would have developed the theory of relativity yourself.

But now let's talk about time in relativity theory. I won't try to answer the ancient question about the true meaning of time. I'm interested only in how we measure it. In your *Principia,* Mr. Newton, you indicated that in a particular reference system a clock can be placed at any point in space and show the same time as all other clocks at all other points in space. In other words: the clocks are synchronized.

If you accept my principle of the constant speed of light, the same synchronization can be realized. If two clocks are in the same place, there is no problem whatsoever. To set our wristwatch by a wall clock, we merely read the time on the clock and set our watch accordingly.

But if the two clocks we want to synchronize are far apart, things aren't so simple. Take two observers, each with a clock. One observer is on Earth, the other on Mars. Light needs some time to reach Mars from Earth, say, five minutes—the exact time, of course, will change with the relative position of the planets.

Let's assume that the observer on Mars wants to set his clock so that it shows the same time as the clock of the Earthbound observer. He sends a radio signal to Earth and asks for the time. The Earthbound observer immediately sends the desired signal, which is registered on Mars a few minutes later. But the observer on Mars won't be able to set his clock exactly because he has to reckon the time it took the light signal to get from Earth to Mars.

NEWTON

The time difference can easily be calculated as long as the exact distance between Earth and Mars is known at that point in time. But to find that distance is not so easy.

EINSTEIN

It certainly isn't. But we don't really need to know the exact distance. There's a trick we can use. Assume the Earthbound observer sends a signal to Mars that bounces off the surface of Mars and is received back on Earth after a given time. I've read that things like this can actually be done these days. Is that true, Haller?

HALLER

Quite true. There are intense and highly focused beams of light that we call laser beams. They can be sent to Mars, and the rebounding signal can be registered on Earth.

EINSTEIN

Excellent. That means we only have to divide the time it takes the signal to travel from Earth to Mars and back by two, and there is our desired difference. Let's assume it's exactly five minutes. At a certain time—say, at eight o'clock sharp, London time—the time signal is sent by a transmitter on Earth. When the signal gets to Mars, the Marsbound observer must set his watch to 8:05. Then the two clocks will be synchronized.

NEWTON

Yes, I can see that clocks can fairly easily be synchronized by means of radio or light signals. Since you can carry out this procedure with any clock in any location in space, I conclude that we can synchronize all of space in this way—or, at least, a large fraction of space. All clocks in this space will then be synchronized and will display the same time. This situation reminds me of absolute time, which I discussed in the *Principia.*

EINSTEIN

We had better leave your absolute time out of our discussion, Newton. There's something important I should add. Synchronizing the two clocks on Earth and on Mars was easy because both were at rest with respect to each other. True, the Earthbound clock moves through space at the speed of Earth, and the Marsbound clock at the speed of Mars. But the difference between the two speeds is very small when compared with the speed of light, not more than 10 kilometers a second, I believe, and so we can ignore it here.

The same is true of synchronizing clocks anywhere in space,

which you were speaking of, as long as the clocks are either at rest with respect to the clock on Earth or moving slowly. But as soon as one of the clocks travels through space at a very great speed, things look quite different. That's why I want to leave absolute time out of our discussion.

NEWTON

No offense, Einstein, but I'd like to know what happens when the clocks are moving with respect to each other.

Einstein took a cigar ceremoniously from its case and lit it with an old-fashioned lighter. Newton, a nonsmoker, couldn't have known that Einstein liked to smoke cigars, and not the best cigars at that. He watched Einstein skeptically, but said nothing. Finally Einstein continued his lecture. He began to talk about time dilation, probably the strangest aspect of relativity theory.

EINSTEIN

Gentlemen, what I wish to explain now can be demonstrated with any clock. For the sake of simplicity, I'll construct a special clock to focus on the salient point.

Let's place a satellite in space exactly 150,000 kilometers away from Earth. It is equipped with a special mirror that can reflect a signal sent from Earth. I think the most suitable signals would be the laser beams mentioned by Haller.

I chose the particular distance between Earth and the satellite because it takes exactly half a second for the light to travel that distance. So to travel from Earth to satellite and back to Earth will take the light exactly one second.

HALLER (interjecting)

At this point I'd like to say that satellites of the kind you have just described actually exist. They are being put to practical use for such things as telephone communications between Europe and California. The telephone signal is sent from a transmitter on Earth to a satellite, and then back to the receiver on Earth. During a phone conversation between London and Los Angeles, the telephone signal travels a distance of about 150,000 kilometers; for that it will obviously need about half a second. You clearly notice this time lag when you talk to someone in California. Because of the unusual time lags in the telephone signals, inexperienced telephone users sometimes get quite unnerved.

Neither Newton nor Einstein had ever telephoned to California. Newton was all set to try. We decided to do a little experiment by putting the telephone in Einstein's apartment to slightly improper use, charging the call to the Albert Einstein Society. At that moment the time in California was shortly after midnight. So I dialed a number that I knew would be answered at all hours. It was the fire station at the California Institute of Technology in Pasadena, where I had worked several times in the past. Someone immediately answered the telephone. I handed the receiver to Newton, and he started a trivial conversation with a Caltech operator, just to find out whether the time lag I had mentioned really existed.

Newton was obviously impressed with our little experiment. This was the first time he had experienced an effect of the finite travel speed of electromagnetic signals.

Our host appeared to have followed Newton's phone experiment with interest.

EINSTEIN

Let's get back to my clock. The light signal needs one second to travel from Earth to the satellite and back to Earth. What I've built here is a rather unusual clock. Its time is defined not by a pendulum or by the oscillation of a quartz crystal but by the bouncing of a light signal back and forth between a satellite and a station on Earth, a bit like a light pendulum. We might call it a light clock.

Let's imagine now that this light clock is being observed from a spacecraft that is moving rapidly past the Earth. What will the observer see from the window of his spacecraft? He will see both the Earth and the satellite moving rapidly past his spacecraft, for he considers himself to be at rest. Now this observer is supposed to be able to follow the light signals as they travel back and forth.

NEWTON

Can that actually be done? I should think it would be hard to follow the light signals. The photons are simply bouncing back and forth between Earth and satellite.

EINSTEIN

In principle, that isn't a problem, and right now I'm interested only in the principle. Suppose, for instance, that our light clock

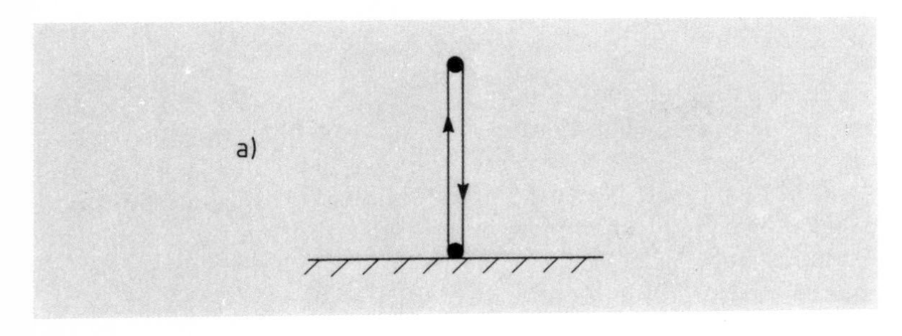

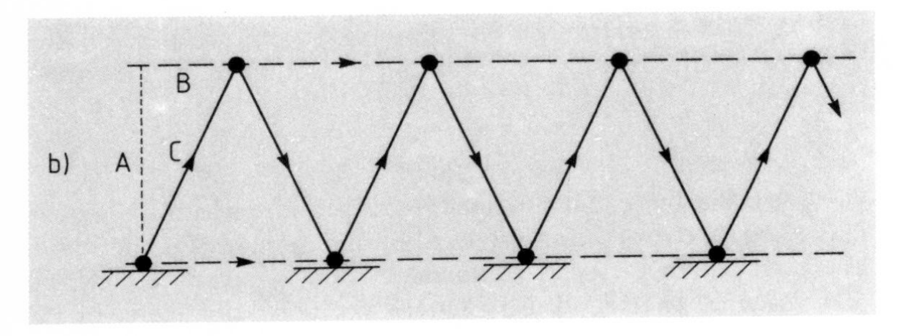

Fig. 9.2 The light clock at rest and in motion. (a) A laser signal is sent from an Earthbound transmitter to a satellite in stationary orbit, from which it is reflected to Earth, bounced back to the satellite, and so forth. (b) Seen from the passing spacecraft, the light signal travels along a zigzag path. The broken lines are the trajectories of the satellite and the transmitter, respectively.

sends out special radio signals every time the light ray bounces and that the spacecraft receives the signal. In this way, the observer in the spacecraft is well informed about the running of the light clock. He can then follow the light signal on its path through space, at least indirectly.

Einstein took a sheet of paper and drew the path of a light signal.

EINSTEIN

While the signal is traveling for one second, both the Earth and the satellite are moving through space; the observer in the spacecraft sees the light signal as a zigzag line. Let's observe the path of a signal during the time in which the signal is making one round trip, from the Earth to the satellite and back.

NEWTON

You mean in one second?

EINSTEIN

I didn't say that, Newton; I was talking only about the time it takes the signal to complete one round-trip from the Earth to the satellite and back. We're going to discover that this time period, when measured by the observer in his spacecraft, is generally not equal to one second.

Puzzled, even alarmed, Newton looked at Einstein. The latter pointed at his drawing and continued, unmoved by Newton's reaction.

EINSTEIN

We see right away that the light signal in the spacecraft's system has a longer path than in the Earthbound system. We know that the path length in the Earthbound system is exactly equal to a light second, that is, approximately 300,000 kilometers. In the spacecraft's system, the exact length of the path depends on the speed of the spacecraft relative to that of the Earth. If the spacecraft's speed is no more, say, than a few kilometers a second, you'll notice almost no change in the length of the path. But if the spacecraft is moving faster relative to the Earth, let's say at 100,000 kilometers a second, the effect will be clearly noticeable.

HALLER

It's not at all unusual for the path in the spacecraft's system to be longer. The same holds true for a passenger on the deck of a boat traveling down a river, when the passenger walks back and forth between port and starboard. For an observer on shore, the passenger's path is much longer than the width of the boat because in the time it takes the passenger to walk across the boat, the boat has moved some distance.

In the system of the observer on shore, it's a zigzag path, like the path of the light signal Einstein has just sketched.

NEWTON (carefully choosing his words)

That is clear. Still, I think there is a difference between Einstein's time clock and your boat example. The speed of the passenger walking on deck obviously depends not only on how fast he is walking but also on how fast the boat is moving. The faster both move, the greater the resulting speed of the passenger.

But with light, things are different, since light has the same speed in every system. So the light that follows Einstein's zigzag line will have that precise speed and not a wink more. And that means . . .

Newton had suddenly stopped talking. His face reflected intense thought. Einstein jumped up and continued where Newton had left off.

EINSTEIN

Exactly. That means that the time in the spacecraft's system runs differently from the time on Earth. The path that the light signal has to travel is longer in the spacecraft's system than in the Earthbound system. On the other hand, the speed is the same in both systems, so the time interval must be greater than a second. In other words: time is being dilated. A second in the Earthbound system—that is, a second for our light clock—appears in our spacecraft's system as an interval longer than a second. The experts call this "time dilation," but it could just as well be called the stretching of time.

As Newton listened, he turned pale. These new realizations, as Einstein had explained them, entered his consciousness almost visibly. I could well imagine how he felt when confronted for the first time with one of the most surprising phenomena of our world: the fact that there is no universal time; that even time depends on the state of motion.

Einstein, who sympathized with Newton, held his tongue. There was a strange silence in Einstein's apartment, each of us following his own thoughts. After a while, Newton resumed the discussion.

NEWTON

I'm beginning to realize, Einstein, that the discovery of time dilation—which is in my opinion an amazing phenomenon—finally puts my idea of absolute time out of business. I believe I now understand this astonishing consequence of the universal speed of light. Only a few aspects are not quite clear to me. I'm sure you can explain them.

EINSTEIN

Fire away, Newton! I'll do my best.

NEWTON

I do understand why you use this complicated image of a light clock, this system of Earth and satellite, to present your thoughts. But are you sure that time dilation, however plausible it looks in the context of your light clock and the constant speed of light, really implies a universal stretching of time? In other words: Is time dilation also true of normal clocks like your wristwatch, or of any other periodical process such as your pulse?

EINSTEIN

Of course. I've used the light clock to derive my notion of time dilation simply because it's an easy way of understanding it. I could have used a normal clock, but that would make it harder to illustrate the effect. Time dilation, however, is true for all clocks. It isn't directly concerned with clocks, but with the flow of time. All events are influenced equally, including chemical and biological processes. And even the aging process.

St. Augustine once wrote: "Time is like a river of passing events, and strong is its current; no sooner is a thing brought to sight than it is swept by and another takes its place." He was right, though he didn't know that the flow of time isn't steady but depends on the state of motion of the observer. If you like the example of the river, don't think of evenly flowing water. Think of a wide stream with varying currents, with rapids, with slow-flowing side arms.

For the observer in a spacecraft, the flow of events on Earth, as he sees it, is slowed down. If, for instance, he could observe the heartbeat of a colleague on Earth with the help of a radio signal, he would find that the heart didn't beat at the normal pulse rate of about 70 beats a minute; it might be down to 30 a minute. How much it would slow down would obviously depend on the speed of the spacecraft. But that doesn't really present a problem; to a rapidly passing observer, all events appear slowed down.

NEWTON (interjecting)

It does sound rather crazy at first glance. There is one point I still don't grasp. You say that events on Earth appear to be slowed down to the observer in a passing spacecraft. Fine, I accept that.

But now I'll turn the tables. I place an observer on Earth and let him view the passing spacecraft. Since it is rushing past him at great speed, he should be able to observe time dilation in all the processes that are taking place on it.

For the Earthbound observer, time would appear to run more slowly in a spacecraft than on Earth. Doesn't that contradict what we've said before? There it was the opposite. For the observer in the spacecraft, all Earthbound processes were slowed down. Isn't there a contradiction?

HALLER

Not at all. Let's forget Earth, the satellite, and the spacecraft, and let's look instead at two spacecraft meeting somewhere in space. Both are in linear and uniform motion with respect to each other. Neither is distinguishable from the other. An observer in one of them will notice that the clock in the other runs more slowly than the clock in his own.

Exactly the same observation is made in the other spacecraft, where an observer will notice a slowing down of the clock in the first craft as compared to his own. For both observers, time dilation takes place. There's no contradiction here; it just means that the flow of time depends on the system. To generalize: the flow of time in a moving system, when observed from a system we define to be at rest, will always appear slowed down.

Einstein had nodded in agreement, but Newton had accepted my answer with a frown. He now paused briefly before resuming.

NEWTON

Fine, let's leave this alone for a while. But I would like to find out to what extent time can actually dilate. We should be able to calculate it as a function of relative speed.

He took a pencil and bent over Einstein's drawing.

NEWTON

I believe I know how to calculate it. The distance between the Earth's surface and the satellite, which we have assumed to be 150,000 kilometers, I'll call *A*. That would be the distance the light travels in a system in which the Earth and the satellite are both at rest.

Seen from the moving spacecraft, the Earth and the satellite will move at a certain speed through space, a speed I'll call *v*. In the time in which the laser beam moves from the Earth to the satellite, covering a distance I'll call *C*, both the transmitter on Earth and the satellite will have covered a distance *B* in their journey through space.

[The reader who might experience difficulties with mathematical equations should skip the following little calculations.]

I noticed that Newton was looking ill at ease.

HALLER

Quite right. These are the three distances we have to consider when we calculate time dilation. The ratio C/A is the quantity that measures the stretching of time. While the light signal takes half a second to travel the distance *A* when seen in the Earth's system, it will take C/A times half a second in the spacecraft's system. Since *C*, being the hypotenuse of a right triangle, must always be greater than the adjacent *A*, it follows that the time dilation factor C/A is always larger than one.

This factor clearly plays an important role and has a technical name—the gamma factor. The Greek letter γ is used to denote the ratio C/A.

$$\gamma = \frac{C}{A} = \frac{\Delta t'}{\Delta t}.$$

In this equation I have denoted a time interval measured in the system of an observer at rest with the symbol Δt, and the analogous dilated time interval measured in a moving system with the symbol $\Delta t'$.

NEWTON (excited)

Of course! We just have to calculate the ratio C/A, your gamma factor, as a function of the speed v.

EINSTEIN (interrupting)

That's easily done. The three distances *A*, *B*, and *C* are not independent of one another because they are the three sides of a triangle. We can use Pythagoras's theorem. The sum of the squares of *A* and *B* equals the square of *C*.

Newton wrote down the equation mentioned by Einstein:

$$A^2 + B^2 = C^2.$$

He then juggled the equation in order to obtain the gamma factor C/A:

$$\left(\frac{A}{C}\right)^2 + \left(\frac{B}{C}\right)^2 = 1,$$

$$\left(\frac{A}{C}\right)^2 = 1 - \left(\frac{B}{C}\right)^2,$$

$$\frac{A}{C} = \sqrt{1 - \left(\frac{B}{C}\right)^2}.$$

NEWTON

On the left-hand side of this equation we now have the inverse of the gamma factor, but expressed in terms of the ratio of B to the hypotenuse C. B and C are the distances traveled by the satellite and the light signal in the same span of time. So the distances should have the same ratio as the speeds, with v the speed of the satellite and c that of the light signal:

$$\frac{B}{C} = \frac{v}{c},$$

$$\frac{A}{C} = \frac{\Delta t}{\Delta t'} = \frac{1}{\gamma} = \sqrt{1 - \left(\frac{v}{c}\right)^2}.$$

Now I have the gamma factor as a function of v or, rather, of the ratio v/c:

$$\gamma = \frac{\Delta t'}{\Delta t} = \frac{1}{\sqrt{1 - \left(\frac{v}{c}\right)^2}}.$$

Quod erat demonstrandum.

Newton stared for a while at the equation he had just derived. It is one of the basic equations of relativity theory.

NEWTON

Now everything falls into place. It is clear that the one quantity that is important for all these considerations is the ratio of the observed speeds to the speed of light. Einstein, you've stressed several times that the effects of relativity theory become significant only when the ratio of the relevant speeds to the speed of light becomes non-negligible. I must say I hadn't expected that the square of v/c would be so important. For all the speeds that we encounter in technical processes, this ratio is extremely small, and so its square will be even smaller. That makes it clear that the effects of relativity theory can be ignored for all everyday processes, since the speeds involved in them can't be remotely compared with the speed of light.

HALLER (interjecting)

Dear Sir Isaac, Einstein and I have repeatedly emphasized that your laws of mechanics have not simply proved wrong in relativity theory. At small speeds, the deviations that we now can see explicitly are extremely small; they are practically negligible. Even for very large speeds—say, 100,000 kilometers a second—the deviation of the gamma factor from one is quite modest. In this case it is about 1.06; it differs from one by only 6%.

Meanwhile, Einstein had used his pocket calculator to find the gamma factors for some specific speeds, and had jotted them down on a piece of paper.

EINSTEIN

Look here, Newton: this little table shows the gamma factors for a few instructive speeds.

EINSTEIN'S TABLE

Object	v (km/s)	v/c	Gamma factor
car	0.03	0.0000001	1
airplane	0.5	0.000002	1
rifle bullet	1	0.000003	1
10% of c	30,000	0.1	1.05
50% of c	150,000	0.5	1.155
90% of c	270,000	0.9	2.294
99% of c	297,000	0.99	7.09
99.9% of c	299,000	0.999	22.4

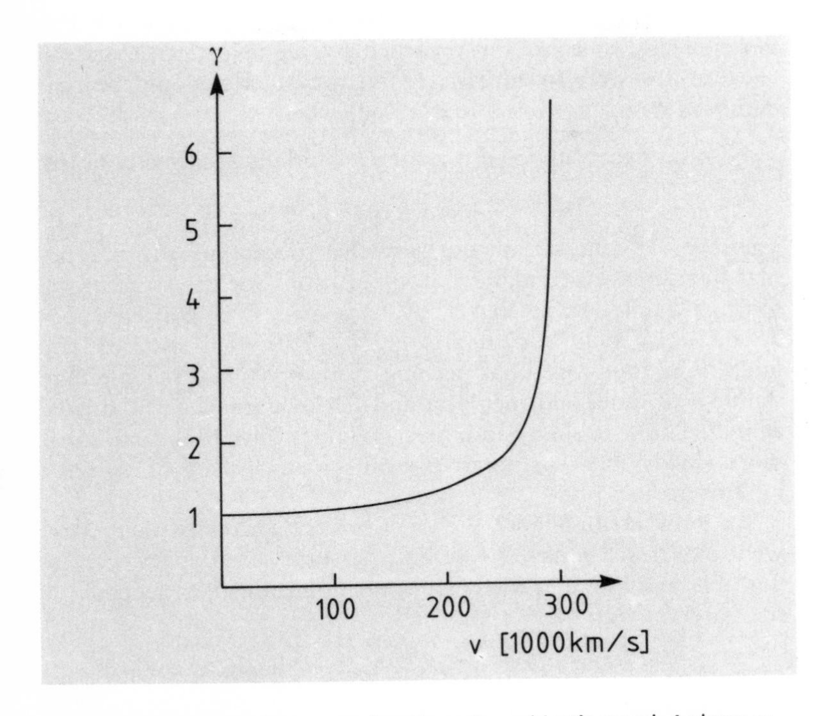

Fig. 9.3 The gamma factor as a function of an object's speed. As long as the gamma factor is small compared with the speed of light, time dilation effects can safely be ignored. They become noticeable only at speeds above 100,000 kilometers a second, and increase as the speed approaches that of light. Note that the latter can never be reached, since that would imply an infinite gamma factor.

As you can see, the gamma factor for normal speeds—I mean speeds we can reasonably imagine—is practically equal to one. Only when speeds approach the speed of light will this factor deviate significantly from one.

As Einstein spoke, I had been sketching a curve that shows the gamma factor as a function of speed. I sketched it from memory since I had often shown it in my university lectures. When I showed it to Newton, he looked at it attentively. Stressing every word carefully, he then spoke.

NEWTON

When we look at the mathematical expression for the gamma factor, we see that this factor becomes greater the closer we get to the speed of light. The latter, however, can never be reached—more precisely, it can't be reached by an object with mass—because if v were to equal c, the gamma factor would become infinitely great.

Newton paced up and down for a few minutes, then continued.

NEWTON

The speed of light is really not like other speeds; it is the quantity that has the most fundamental implications for the structure of space and time, or, better yet, of space-time. Even concepts such as time itself fade in comparison. After all, what is time? You might think our world has nothing more stable and reliable than the flow of time; and then you find out that time can be dilated at will, like a rubber band. Just imagine, time runs four times more slowly in a spacecraft passing us at a speed of 290,000 kilometers a second.

But how about space? We've not really addressed that. After what I've heard today, I wouldn't be surprised if space were as unstable as time, if it too were a phenomenon that depends on the observer.

EINSTEIN

Right, Newton, your conjecture about space will turn out to be true. But we'll talk about that another time. For all that time dilation, I notice that our own time here has passed quickly. It's after one o'clock, and lunch time. I suggest that we close for the morning and turn our attention to a concrete problem—finding a good meal.

Newton and I agreed. We lunched as usual at the Aarbergerhof. After that, we decided to take advantage of the beautiful weather to take the outing we had planned some days before. I invited Einstein and Newton for an afternoon trip by car to Lake Thun and Lake Brienz. After an enjoyable ride and a long walk along the shore of Lake Brienz, we returned to Bern in the evening, happy and relaxed.

During the afternoon, I had secretly been watching Newton. He had been unusually quiet. Although he clearly enjoyed the mountain scenery, his thoughts were still caught up in this morning's discussions. Newton's solid view of this world had started to waver—worse, it was showing cracks. It was a good thing that we had paused for the afternoon. Newton had time to digest the new situation.

I couldn't help thinking about the time when, as a sixteen-year-old high school student, I became acquainted with the basic ideas of relativity theory. At that time I felt like a mountain climber who had started out in good visibility and good spirits, and then suddenly found himself in heavy clouds, unsure of his path. It may take hours of fruitless effort to find the way, finally to emerge above the cloud cover, and then to stand in the midst of splendid mountain scenery under a brilliant sun. Now the climber can continue on his way.

Newton was still climbing through the fog. But I was sure he would soon emerge from the clouds. He would soon have a full view of that panorama of space and time that Einstein had first discovered in 1905.

TEN

Fast Muons Live Longer

The next morning I went directly from my house to Kramgasse. Einstein was already there, and a few minutes later Newton was climbing up the spiral staircase.

Surprisingly, he appeared in good spirits and greeted Einstein cheerfully.

NEWTON

My dear Einstein, you may be looking at a man who slept poorly, but that man can now say of himself that he has understood at least the basic ideas of your theory. So, let's get down to business in our morning session. I must admit that a number of things are still bothering me.

We settled as comfortably in Einstein's living room as the Spartan furnishings would allow.

NEWTON

Beyond any doubt, the meeting we had yesterday was one of the most interesting scientific sessions I have ever attended. I find it astonishing how you started from a simple principle—the universality of the speed of light, which meanwhile has been demonstrated experimentally—to erect the whole structure of your relativity theory. The central point of yesterday's discussion, time dilation, is one of the most important parts of this structure—perhaps *the* most important.

But physics is an experimental science. The finest theoretical structure will come crashing down if the outcome of a single experiment doesn't confirm the expectation. I ask both of you now: What experimental tests of the theory of relativity are there today? Has the theory—which isn't in direct contradiction to my mechanics but is more or less an expansion of it in the case of very rapidly moving objects—been fully proved by experiment? I would be only too happy, my dear Einstein, if that were so, since I now see no other way of solving the problem presented by the universality of the speed of light.

EINSTEIN

My first paper on relativity theory was, of course, very speculative. It soon turned out that my ideas could be developed free of contradictions. It was possible to define a new, relativistic version of mechanics, which then gave us a consistent picture of the dynamics of rapidly moving objects.

I am not certain about the latest experimental tests. Since I've been back in Bern I've been trying to update my information, but I've had too little time to get fully caught up. As far as I know, no experiment contradicts relativity theory. But we have an expert here. Haller, you are a member of our Olympian Academy. I call on you to help us.

HALLER

First of all, I should emphasize, Sir Isaac, that so far we have discussed only a few aspects of Einstein's theory. One of these is time dilation. A more important aspect of the theory of relativity, in my opinion, is still to be dealt with, and we'll talk about that later. For now, however, I'll just comment briefly on research on time dilation.

How should we measure time dilation? In principle, we could simply observe the running of a clock in a rapidly moving vehicle, such as a rocket. But how do we obtain the exceedingly high speeds that we need? To get a measurable effect, the speed of the vehicle should be close to that of light. Even today, with all the technical means at our disposal, it's impossible to accelerate a macroscopic object to such high speeds.

EINSTEIN

The effect of time dilation does exist, of course, in all moving clocks; but at relatively low speeds it's very modest indeed. If extremely accurate clocks were available, we could probably measure this effect even at the relatively moderate speeds reached by rockets today. How about that?

HALLER

Whether time dilation can be measured at moderate speeds—say, some tens or hundreds of kilometers per second—really depends on the precision of the clocks we would use for the test. For now, let's skip the technical problems this raises—but I promise you we'll get back to them. Let's consider how we can measure time dilation in objects moving almost at the speed of light.

Although there is no way of accelerating macroscopic objects to such high speeds in normal labs, it can be done with microscopically small objects like protons and electrons. You don't even have to go into a lab; nature provides us with plenty of rapidly moving particles.

NEWTON

Well, let's assume we observe a rapidly moving particle—your electron, for instance. How are you going to measure time in the system of that electron? You can hardly strap a watch around its neck.

HALLER

And you don't need to. There's a trick for avoiding that problem. We don't use an electron but a particle with a built-in clock. Let's assume we are looking at a particle that will decay by some inherent process, like weak nuclear interaction, after exactly one second.

EINSTEIN (skeptically)

Do particles like that exist in our universe?

HALLER

There are many particles in nature that are not stable but decay shortly after their creation. True, there is no particle with a lifetime of exactly one second—it's the principle that's important here.

Let's now look at a one-second particle, and let it be at rest with respect to us. What happens is clear: after exactly one second, the particle will decay and generate several secondary particles—never mind the details of the decay process.

Now consider another particle of the same kind, moving at speed v relative to us. Start with a small value for v, but increase it gradually. As long as v is small relative to the speed of light, we'll still see the particle decay after one second.

This is easy to measure. We only have to follow its path from creation to decay. The length of this path is simply speed v multiplied by the lifetime of the particle, one second. If the particle has a speed of 1 kilometer a second, it'll travel exactly 1 kilometer and then decay.

NEWTON

Excuse me, Haller, for interrupting—but I believe I know what you're trying to say. If we accelerate the speed of the particle, let's say to 100,000 kilometers a second or more, time dilation will set in, the gamma factor will become important. The particle will then appear to us to live longer than one second. To the observer at rest, its lifetime will be one second multiplied by the gamma factor.

HALLER

Exactly. The length of the path of a particle from the point at which it is created to the point at which it decays is not given merely by its speed but by its speed multiplied by the gamma factor. Here's an example: let's assume the particle moves at the very high speed of 99% of the speed of light, that is, at about 297,000 kilometers a second. If there were no time dilation, the particle would travel through space for 297,000 kilometers and then decay. Its gamma factor for this speed is appreciable—7, to be precise. Which means that the particle moves seven times farther into space than 297,000 kilometers, namely, about 2 million kilometers. That is an immense difference, of course. It could not be missed in an experiment.

EINSTEIN

I admit that this is a very interesting test, but you told us before there is no such thing as a one-second particle in nature. So your test can't really be done this way. How should we go about it?

HALLER

Just a little differently, that's all. Today we know of the existence of a number of particles that can actually be used for experiments of this kind. I'll mention only the best-known and most impressive of these tests, which uses muons. Muons are particles very much like electrons—you might regard them as heavier brothers of the electron. A muon's mass is about 200 times as large as the mass of an electron.

EINSTEIN

How strange! They sound like unusual particles. Do you have any idea why they exist?

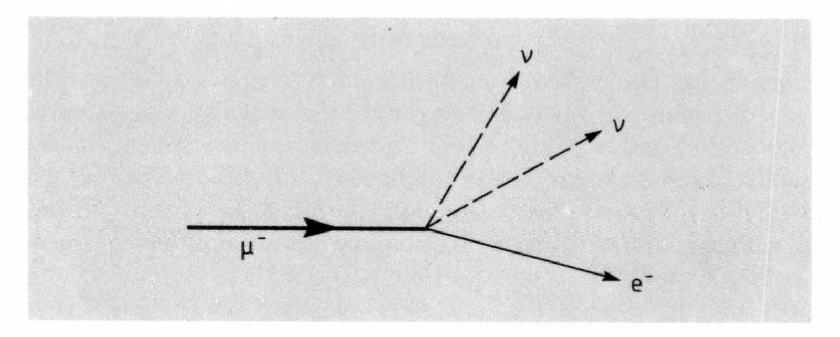

Fig. 10.1 The decay of a muon: here, a negatively charged muon decays into an electron and two neutrinos. Neutrinos are electrically neutral partners of electrons and of muons. It is possible that they are massless like photons, but this has not been verified experimentally.

HALLER

Nobody knows. They seem to have no more use than weeds in a garden. And they are unstable too, so they seem to have no significance for the structure of stable matter in the universe.

NEWTON

Maybe Einstein especially ordered these muons from the creator of our universe so that he could give us an impressive proof of his relativity theory. They are not quite useless if, in fact, we can measure time dilation with their help.

EINSTEIN

If I really had ordered one thing, Newton, it would have been Haller's one-second particle, for the sake of convenience. But no more kidding, gentlemen! Tell us how muons decay.

I produced a graphic design of a muon's decay.

HALLER

Muons were detected in 1937 by an experiment that was designed for the study of cosmic rays.

NEWTON

Excuse me, but what exactly are cosmic rays?

HALLER

The universe is constantly traversed by fast-moving particles. Most of these are protons, the nuclei of the hydrogen atom, but some are nuclei of other elements—helium, for example, or heavy elements such as carbon or iron.

All these particles usually move at a speed close to that of light. When they collide with atomic nuclei in the upper atmosphere, a little explosion occurs. It's caused by a particle reaction that is often quite complicated, so I won't go into it now. In any case, muons are created in these reactions. They fly away from the collision point at almost the speed of light, and many reach the surface of the Earth. Our bodies are constantly bombarded by these particles. Individual atomic nuclei in our bodies are often hit and destroyed by them.

EINSTEIN (having studied my sketch)

According to your sketch, three particles remain after the decay.

HALLER

Yes, an electron, which takes over the electrical charge of the muon, and two other particles that are electrically neutral, called neutrinos. Neutrinos are very difficult to observe since they have practically no interaction with matter—including the matter of which our detectors are made.

That's why it took such a long time before the details of muon decay could be explained in the early sixties. The details are irrelevant for our purpose. What matters is that we can observe the decay of a muon, and we do so by observing the electron that emerges from it.

Now to the question of what the muon's lifetime is. It has been shown that we can't establish a particular time after which a muon will certainly have decayed. We can only establish a probability for the decay of a muon—or, rather, of many muons. Let's assume we track one thousand muons that are at rest and that were created at the same moment. We shall find that after the relatively short time span of only 1.5 microseconds—that is, 1.5 millionths of a second—exactly one-half the particles, 500 of them, have decayed. After another 1.5 microseconds, one-half of the remaining muons, 250, will have decayed. This is called the lifetime of the muon.

If you know what the lifetime of the muon is, it's easy to state the probability of the muon's survival as a function of the time that has elapsed since its creation. The longer the time, the smaller the probability.

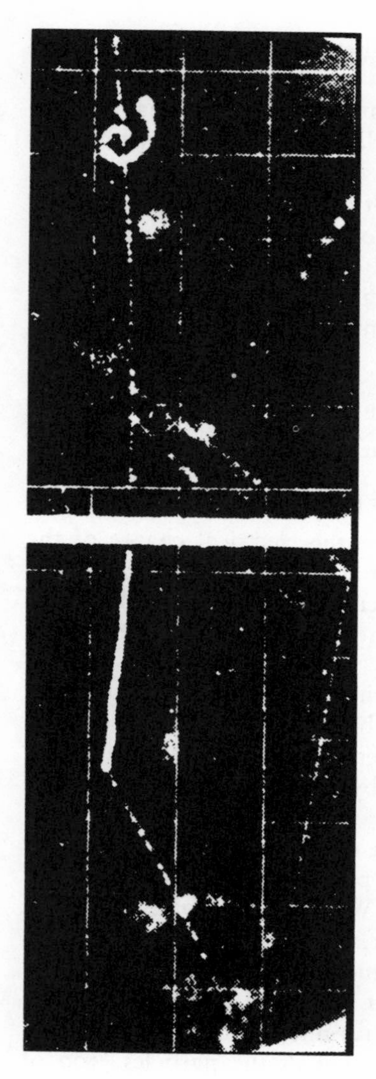

Fig. 10.2 The decay of a muon: here, a positively charged muon penetrates from above into a detector called a cloud chamber. Its path is made visible by the cloud chamber in the form of a track formed by tiny droplets, analogous to the tracks that form behind jet planes in the atmosphere. The muon traverses an aluminum plate; in the process, it is slowed down and decays into a positron (weak track) and two neutrinos (not visible here). The positron is the electrically positive analogue of the electron, or, more accurately, the antiparticle of the electron.

I took a piece of paper and drew an approximate picture of this function.

HALLER

You can never say with absolute certainty that all the muons have decayed after a certain time. Even after a full hour, which is a very long time indeed when compared with the lifetime of a muon,

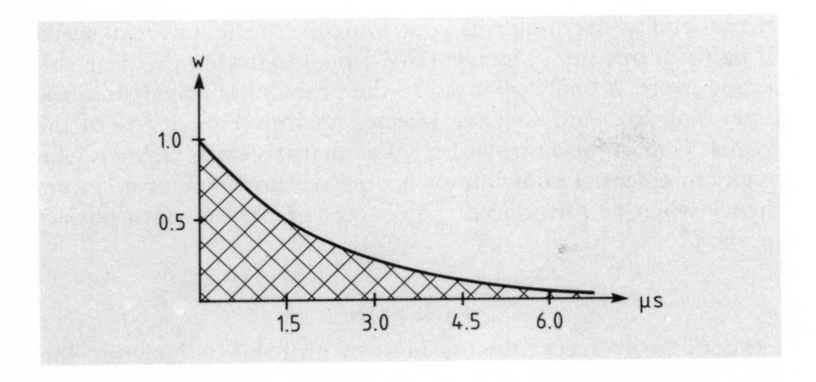

Fig. 10.3 The probability of a muon's decay is plotted here as a function of time: within 1.5 microseconds, the probability that the muon has survived is seen to have dropped to 50% or 0.5 ms. This period is called the lifetime (or half-life) of the muon. After another 1.5 microseconds, the survival probability has decreased by an additional factor of 0.5, down to 0.25. After only 10 microseconds, the probability has become very small indeed, on the order of 1%.

there is still a finite, if minute, probability that one of the muons, a real Methuselah, may have survived.

EINSTEIN (impatiently)

Enough, enough, Haller! Newton and I do understand what is meant by probability. I just don't understand why there should be any probability at all. Why isn't there one and only one lifetime for muons? I do see a problem here. I don't understand why muons would throw dice before decaying. Why don't they just decay after the expiration of their life span, let's say after 1.5 microseconds, and let that be the end of it—the same as with your hypothetical one-second particle?

Newton smiled at these last words of Einstein's.

NEWTON

Mr. Einstein, a few days ago I hadn't the slightest idea of relativity theory; I used my time in Cambridge to familiarize myself a little with atomic physics, and I learned that all atomic processes can be described only in terms of probabilities. That's a basic principle of atomic theory, or, if you wish, of quantum theory. But that theory was developed in the 1920s; it is probably just as fundamental and as irrefutable as your principle of the universal speed of light. In one book, incidentally, I read that you, my dear colleague, were strongly opposed to the probability principle at the time. You are said to have rejected quantum theory with the words: God doesn't throw dice. Was that the same Einstein who made an essential contribution to the development of quantum theory when he introduced the concept of photons into physics in 1905?

EINSTEIN

I cannot easily accept this business of probability, Newton. But since you have quoted the sentence about God that I'm supposed to have said—and I have no qualms about that—I will now add another: The muon may decay whenever and however it pleases, but it won't throw dice. The laws of nature are clear and unequivocal, not fuzzy. Maybe our present-day understanding of these matters doesn't permit us to do more than state probabilities of muon decay, but someday we will surely be able to make exact statements about the decay of every individual muon.

Einstein's last remarks made me more and more uncomfortable. The discussion couldn't go on like this without the risk of the three of us getting into a violent argument about the foundations of quantum theory. After all, throughout the last decades of his life, Einstein had argued against the interpretation of quantum theory that is generally accepted today. His disputes with Niels Bohr, one of the founders of atomic theory, are today part of the history of physics. I attempted to reorient the discussion.

HALLER

Gentlemen, I mentioned the decay of muons just to give an example of time dilation. The last thing I wanted to start was an argument about quantum theory. I suggest we leave atomic physics and quantum theory alone and get back to where we started. The only thing we need of muons is their lifetime. Whether the probability statement about their decay reflects a deep property of the quantum process or merely our ignorance of the details of what happens inside a muon, we'll just leave open.

Newton expressed agreement, and Einstein too nodded in resigned assent.

NEWTON

But I do hope we'll get back some other time to the problem of probability in quantum theory. I'm really more interested in that than in the questions of space and time in relativity theory.

HALLER

Back to muons, then. If we examine the decay of many muons, we can easily determine their lifetime; it's the very short time span of 1.5 microseconds that I mentioned earlier. And that means the muon is a particle that can be used as a clock. At the very least, we can use it to measure 1.5 microseconds. To a particle physicist, incidentally, that is not a particularly short time, quite to the contrary. Light travels almost half a kilometer in that time. Today, we are capable of measuring times that are many orders of magnitude smaller than the muon's lifetime.

NEWTON

Just a second! Didn't you say earlier that here on the Earth's surface we can find many muons that were produced by cosmic rays striking our planet—or, more precisely, striking the upper layers of our atmosphere? You also said that muons move almost as fast as light. We would therefore expect muons to travel about half a kilometer through space before decaying. Maybe some of them manage 2 or 3 kilometers. But on the average, their range should not exceed 1 kilometer. Now the Earth's atmosphere is considerably thicker than 1 kilometer; it's more like 30 kilometers, so how do those muons make it all the way to the Earth's surface? Wouldn't we expect virtually all muons to decay much earlier, on their way through the upper layers of the atmosphere?

Newton leaped out of his chair, paced the floor for a few moments, and then tapped Einstein on the shoulder.

NEWTON

Mr. Einstein, I believe you win. The muons here on the Earth's surface provide the proof. Of course they make it this far with no problem—it's time dilation that makes them go a lot farther than we would naively expect from their lifetime. That's analogous to the hypothetical particles that Haller discussed earlier.

EINSTEIN (turning to me)

What about a quantitative proof of time dilation? Just by observing a few muons here on the Earth's surface we can't possibly conclude that relativity theory is correct. Though I must admit that things look good for my cause, or, rather, for relativity theory.

HALLER

Nevertheless, Newton's conclusions are justified. Without time dilation things look bad for the muons. They would have little chance of reaching the Earth's surface. We know that most muons are produced about 15 kilometers above sea level. Without time dilation, half of them would vanish after moving half a kilometer. It's easy to reckon that only a tiny fraction of the muons—about one-billionth—would reach the ground. Chances are poor for a muon to be found close to the surface of the Earth. Yet there are many of them. The only way we can explain it is to assume that the fast-moving muon that a researcher detects on the surface of the Earth ages less rapidly than a muon at rest or in slow motion. Time dilation is the only generally accepted explanation of this phenomenon.

EINSTEIN (not giving up)

That does sound pretty convincing. But time dilation in a moving system isn't predicted by relativity theory alone. After all, you can calculate the dilation effect precisely as a function of speed or of the gamma factor. But how about a quantitative test, let's say by means of muon decay? I'm afraid the muons created by cosmic radiation may not be very well suited for that.

HALLER

Unfortunately, I must agree. It's a good thing we don't have to rely on cosmic muons. There are a lot of labs for nuclear and particle physics where we can produce intense beams of muons. Of course, these beams are not produced solely to test relativity theory. No serious physicist doubts the validity of the theory. We use muons for other purposes—for instance, to determine the interior structure of atomic nuclei.

In 1976, an experiment was carried out at CERN to test in detail the predictions of relativity theory. Muons created by means of particle collisions were immediately introduced into a

Fig. 10.4 The storage ring at CERN, used for a detailed investigation of time dilation. The muons are stored in a ring-shaped vacuum tube surrounded by particle counters. These counters register the decaying muons by means of the electrons emitted in the process.
The verification of time dilation was a by-product of this experiment, whose main purpose was precise measurement of the magnetic properties of the muon. (Courtesy of CERN.)

ring-shaped vacuum tube. They moved comparatively fast, at about 99.94% of the speed of light.

The main advantage of these muons over cosmic muons is that we have precise knowledge of where and when they are created and how fast they move. That gave us the prerequisites for a precise quantitative test of time dilation and thereby of relativity theory. The magnetic field guaranteed that the muons would move at a constant speed inside the ring. They were essentially stored there, which is why we call the device a storage ring.

NEWTON

But in that experiment, how do you determine the time at which a muon decays in the ring?

HALLER

That's no problem. You surround the storage ring with particle detectors that are capable of registering the electrons produced by

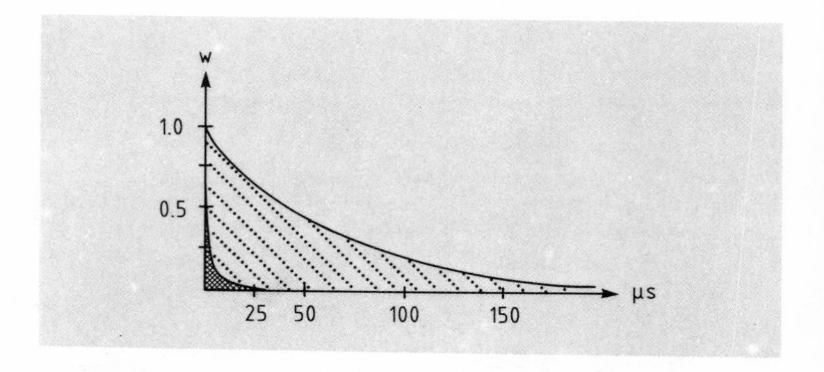

Fig. 10.5 The probability of muon decay as measured at CERN (shaded area), corresponding to a lifetime of 44 microseconds. For comparison, the crossed area denotes the distribution of muons at rest. This would have been the observation if there were no such phenomenon as time dilation. The CERN experiments yield a gamma factor of about 29, in excellent agreement with the predictions of relativity theory.

the decaying muons. Practically all of these electrons are emitted sideways during decay. They leave the storage ring and always fly through one of the particle counters.

EINSTEIN

What a fascinating way to measure time dilation! You simply register electrons as a function of time, and you automatically find how many muons decay per unit time. If you have a sufficient number of muons in the storage ring, you can certainly measure time dilation quite accurately. Well, Haller, don't keep us waiting: What did they find at CERN?

HALLER

I anticipated the result when I said that we have no reason today to have any doubts about relativity theory. But let's get down to hard facts. The CERN experiment showed the lifetime of muons to be 44 microseconds—thirty times as long as that of a muon at rest.

NEWTON

Wait a moment! That factor of approximately 30 would be the gamma factor. Let's check whether that's correct. The speed of the muons in the ring, you said, was 0.9994 times the speed of light. With that we have a gamma factor of

$$\gamma = \frac{1}{\sqrt{1 - \left(\frac{v}{c}\right)^2}} = \frac{1}{\sqrt{1 - (0.9994)^2}} = 28.9.$$

Gentlemen, what do you say to that?

It had taken Newton only a few moments to calculate this result.

HALLER

The measured results correspond precisely to the prediction of relativity theory, and that to within about 0.2%.

Newton looked at Einstein, who was gazing happily out of the window.

NEWTON

What would you do, Einstein, if the physicists at CERN had observed a deviation from relativity theory?

EINSTEIN

I'd like to leave that question unanswered, Newton. You yourself said a while ago that you saw no other solution to the problem of the constant, universal speed of light. Relativity theory simply had to be correct. I would have felt very sorry for our good Lord if he hadn't thought up this solution.

Coming from Einstein, this remark sounded amusing and not at all arrogant, and all three members of the Olympian Academy burst out laughing. A few minutes later, an unsuspecting passerby in Kramgasse might have seen three gentlemen in animated conversation coming out of No. 49 and heading for an inn near Bärenplatz for the approaching noon hour.

ELEVEN

The Twin Paradox

Around two o'clock we returned to Einstein's apartment. During lunch I had deliberately avoided talking about relativity theory. Instead, we talked about various methods of detecting particles. The discussion during lunch and the walk along the Aare river afterward concerned Geiger counters, cloud chambers, bubble chambers, spark chambers, wire chambers, and other such apparatus that can help us find the traces left by passing elementary particles. All these details of elementary particle physics would be of little interest to the nonphysicists.

Upon our arrival at Einstein's apartment, Newton led the discussion back to time dilation.

NEWTON

We know that time dilation affects all moving systems, so we should be able to detect it by placing high-precision clocks in, say, an airplane or an automobile. How about that?

EINSTEIN

O.K., let's perform a thought experiment to find out. Assume we take a high-precision clock and drive in a car at a more or less constant speed of 120 kilometers an hour from Bern to Zurich and back. That will take about two hours.

Einstein took a pencil and did a brief calculation. He soon had the result on paper.

EINSTEIN

The *v/c* ratio is minuscule here, on the order of 10^{-7}. That means the gamma factor differs from one by only 6×10^{-15}. We will be driving for two hours, or 7,200 seconds; the effect of time dilation would ultimately stretch our time by 6×10^{-15}, multiplied by 7,200 seconds, making it a little less than 10^{-10} seconds. When we return to Bern and compare our clock with one that has remained at rest in that city, we should observe a time difference of some 6×10^{-11} seconds.

HALLER

I'm afraid I have to disappoint you. Such a slight time difference can't be measured with the clocks we have today. As I said before, the best atomic clocks reach a precision of one part in 10^{14}, and that's a little less than the precision demanded in the case we have just discussed.

To make the effect even observable, we would have to increase the speed of the car by a factor of 10 and race to Zurich at 1,200 kilometers an hour. No car will go that fast. Besides, we would be exceeding the Swiss speed limit, which is 120 kilometers an hour, by more than 1,000.

EINSTEIN

How about an airplane, which to my knowledge can easily go faster than 1,000 kilometers an hour?

HALLER

Fast jet planes have certainly been used for experiments of this kind. Let's assume we fly once around the Earth in a jet plane at 1,000 kilometers an hour; the total flight time will be 36 hours. The gamma factor's deviation from one amounts in this case to 0.5×10^{-12}. The effect of time dilation on the full flight time is this number multiplied by $36 \times 3{,}600$ seconds, which makes 10^{-7} seconds, a time difference that can easily be measured.

In the early 1970s, scientists at the U.S. Naval Observatory in Washington carried out a simple experiment in which a physicist flew with a scheduled airline once around the globe. He traveled with several atomic clocks placed on the seat next to him. After his return to Washington, his clocks were compared with similar clocks that had remained in Washington. And true enough, the traveling clocks lagged slightly behind the stationary clocks. The result was in perfect agreement with Einstein's theory. This was an inexpensive experiment when compared with the CERN muon experiment, by the way. It cost only the two air fares, one for the physicist and one for his clocks.

Newton had been paying close attention.

NEWTON

That is fine, Einstein. I don't think we need further proof of time dilation. I'm fully convinced now that there is no absolute time independent of the observer. Still, what a crazy phenomenon time is! Clocks run more slowly when they are in motion.

Shaking his head, Newton got up and walked toward the corner of the room, where an old grandfather clock stood. Its hands clearly hadn't moved for a long time. He stared at its face for a while and then pulled up the weight. The ticking of the clock was now clearly audible in the quiet room.

NEWTON

And still we don't know what time is really all about. Suppose I were to remove this clock from the room, along with all other clocks, including all atoms, which are nothing but clocks either . . . Suppose I were then faced with an empty room: Would I have time at all, would there be a flow of time? And if there were, what is it that would be flowing? Does time exist without matter? When will we finally have the answer to all this?

Newton walked to the window. He was lost in thought, looking out over Kramgasse. I suggested a tea break. It was Einstein's turn to make the tea. Newton and I just sat there, following our own thoughts.

NEWTON

Last night, shortly before going to bed, I took a walk to the university. When I looked up at the firmament, it occurred to me that we could use the effect of time dilation, with the aid of rapidly moving rockets, to explore our galaxy and perhaps even other galaxies.

A few days ago in Cambridge, I read that light takes about thirty thousand years to travel from Earth to the stars in the center of our Milky Way. We've been naïve enough to believe that human beings, who seldom live longer than a hundred years, could never undertake that kind of voyage. But this is not so, as we have learned from the muons. Time dilation helps, or, rather, it could help if we were able to build spacecraft capable of reaching speeds close to, if not exactly equal to, the speed of light. What amazing possibilities that would open up! Man, that Earthbound creature, would be able to explore distant space. What do you think? Does that sound realistic?

HALLER

You're right in principle, Sir Isaac. But unfortunately only in principle. You have to remember that the great effect of time dilation

Fig. 11.1 The Andromeda galaxy, the galaxy that is closest to our own, about 2.5 million light years from Earth. Consisting of some 200 billion stars, it is about twice the size of the Milky Way. There is every reason to believe thousands of solar systems exist much like our own. If a spacecraft could travel at a speed close to that of light, a visit to Andromeda should be feasible in principle. When returning to Earth, however, the astronauts would find that about 5 million years had passed on their home planet.

in the case of the muons, where we had a time-stretching gamma factor of 30, was reached only because these particles moved through space at more than 99% of the speed of light. With today's technology it is impossible to accelerate a macroscopic object, whether a bullet or a rocket, to a speed close to that of light. But let's leave this technical problem alone for the time being.

For a man to fly from Earth to the center of the galaxy in thirty years, the corresponding gamma factor would have to be 30,000/30, which makes 1,000. Without the gamma factor, that is, without time dilation, the whole enterprise would be hopeless. During his thirty-year voyage at a speed close to that of light, the astronaut would travel a distance a little less than what light would travel in the same period. He would go just a little farther than the distance from Earth to the fixed stars surrounding our sun. It's easy to determine the speed that corresponds to a gamma factor of 1,000.

At that moment Einstein came in with the tea. He had obviously overheard our conversation.

EINSTEIN

You really haven't changed, Sir Isaac; you're still looking at the heavens. As far as I'm concerned, I admit that I feel quite comfortable here on our little old planet. As a companion of our sun, it moves through space pretty fast too. Can you imagine anything better than spacecraft Earth—a vehicle with lakes and green forests, with cities like Bern? Very well, Newton, let's see how close to the speed of light you would have to travel in order to reach the center of the galaxy within thirty years.

By this time, I had finished my little calculation. The gamma factor is given by the formula

$$\gamma = \frac{1}{\sqrt{1 - \left(\frac{v}{c}\right)^2}}$$

so we can calculate the ratio of our spacecraft's speed to the speed of light c by

$$\frac{v}{c} = \sqrt{1 - \frac{1}{\gamma^2}}.$$

When gamma = 1,000 we get

$$\frac{v}{c} = \sqrt{1 - (10^{-7})} = 0.9999995\,.$$

HALLER

You see, Sir Isaac, the space traveler needs to fly at a speed practically equal to that of light. In a moment we'll find out that it would take an enormous amount of energy to accelerate a spacecraft to a speed of that order. It can't be done with today's technology, and I don't think it will become feasible in the near future.

NEWTON

All right, Haller. I realize that a voyage to the center of the galaxy, or merely to the neighboring fixed stars, belongs to the realm of fantasy for now. Still, even if we can't take that trip ourselves, we can do so in our thoughts. As Einstein likes to say: We can perform a thought experiment.

I was also thinking last night that it should be possible to go to some star in a rapid spacecraft and then back to Earth, and to discover that more time had elapsed on Earth than in the spacecraft.

EINSTEIN

There's nothing wrong with that. Let's assume a space traveler leaves Earth in a spacecraft at a speed of 260,000 kilometers a second. That speed corresponds fairly closely to a gamma factor of 2. Let's further assume that this space traveler leaves his twin brother behind on Earth. He is 30 years old when he leaves. He travels away from Earth for 10 years, then puts on the brakes, turns around, and heads back to Earth on the shortest trajectory. After 10 more years he arrives back on Earth, now 50 years old. On his arrival, he'll see that his twin brother has aged 40 years in the meantime. The twin has just celebrated his 70th birthday.

NEWTON

You've just described a fascinating application of the theory of relativity, Einstein. I must admit I had not thought about processes of life and of human aging in connection with time dilation. But sure enough: if clocks tick more slowly when in motion, if time is expanded by motion, then processes of human life such as biological aging will also slow down.

HALLER

Careful, Sir Isaac. The way you just expressed it, one might think time dilation could be used as a fountain of youth, to play a trick on the aging process. That, of course, is impossible; time dilation, after all, is an apparent effect perceived by the observer who remains Earthbound. In the spacecraft, or rather in the reference frame that moves with it, all processes, including chemical and biological ones in the astronaut's body, happen at their typical rate just as they would do on Earth. It is just to the observer on Earth that these processes would seem to have slowed down if he had the chance to watch them over a large and constantly changing distance. The heart of a space traveler, for example, beats 60 times a minute. If his twin brother on Earth were to register it by means of appropriate radio signals, he would get a reading of 30 a minute. Similarly, currents in the brain, in other words, the astronaut's thought processes, would equally appear to have slowed down.

Time dilation, then, can't help us gain additional time of a useful, livable nature. When the space traveler finally returns to Earth and finds himself younger by 20 years than his twin brother, he notices that from the time he left he has lived only half as much as his brother, with half the thinking, half the food, half the drink, and half the sleep.

EINSTEIN

I'm afraid he's experienced even less. It must have been horrible to be locked up in a small spacecraft for 20 years. If it were up to me, I would stick to our pleasant Earthbound existence. If I had a twin brother, I would send *him* on the trip.

NEWTON

So we can't create a fountain of youth with the help of Einstein's time dilation. But another problem has occurred to me. Let's look at those twin brothers again. One of them stays on Earth, the other moves away from Earth at a constant speed and then, after turning around, moves toward Earth with the same speed. Both twins are located in an inertial system.

As you just mentioned, Haller, the Earthbound twin would have no problem observing time dilation in his spacebound brother if he were capable of seeing his brother's vital processes. But there is no basic difference between the twins: if the astronaut were to look back on his brother on Earth, he would notice the identical time dilation in him; both his brother and the Earth are moving through space when viewed from his inertial system. That would make him expect that when he gets back to Earth, it will be he himself rather than his Earthbound twin who has aged. I must admit the whole thing looks suspicious to me—it is a real twin paradox.

EINSTEIN (sipping his tea)

I've been waiting for that objection. On one point you're right: if the twins moved uniformly with respect to each other, each in his own inertial system, there wouldn't be any reason to prefer one to the other. Both twins observe time dilation when they consider the space-traveling brother in motion. But things are not quite as democratic as that. There is one important difference between the twin on Earth and the twin in space: the space traveler moves away from Earth and comes back after a given time, which means he can't have been moving uniformly through space along a straight line. At some point he has to decelerate the spacecraft, turn it around, and accelerate in the opposite direction. While he is turning, he is not in an inertial system. That means our observation of time dilation, especially the calculation of the gamma factor, doesn't apply to the entire motion of the space traveler; but it does apply to the motion of the brother on Earth.

This is by no means a symmetrical situation. The space traveler is at a disadvantage; unlike his Earthbound twin he has to go through this whole sequence of slowing down, turning around, and accelerating again. That's why it's the Earthbound twin who winds up twenty years his senior.

NEWTON

I concede that there is a difference between the twins, as you just said. But let's clarify the matter by representing the twins' journey through space and time by means of their respective appropriate world lines.

He took pencil and paper and sketched a space-time diagram with world lines for the twins (see fig. 11.2).

NEWTON

The difference between the twins becomes clear when you look at these world lines. The world line of the twin at rest is a straight line; that of his space traveler twin is obviously not a straight line. Depending on the various accelerations or decelerations, it becomes a fairly complicated curve, which will meet the world line of his Earthbound brother at the end of his trip.

EINSTEIN

You were right when you drew the traveler's turnaround as a gradual process rather than an abrupt reversal. We should also

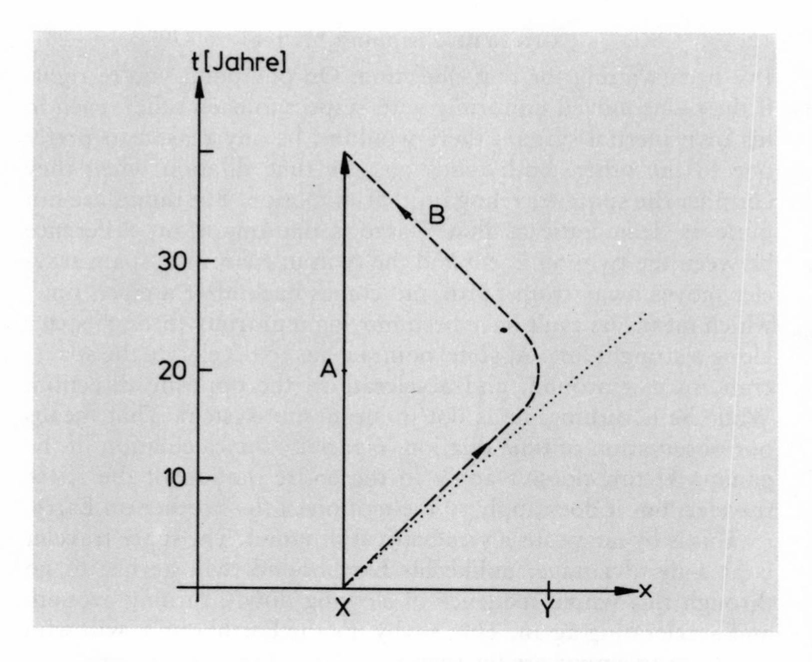

Fig. 11.2 Space-time diagram for the twin paradox: the world line of one twin, who remains at point *X*, runs parallel to the time axis; the other twin travels from *X* to a distant point *Y*, turns around, and travels back to *X*. The dotted line shows the path of a light signal emitted from *X* at the time of the second twin's departure. His travel runs almost parallel to the light signal's path: since he is traveling at almost the speed of light (in the present case, at 260,000 kilometers a second), his path proceeds almost parallel to the light cone.

bear in mind that the traveler doesn't start out at full speed but has to do a lot of accelerating before he reaches that speed. But let's ignore such details for the moment.

NEWTON

I doubt that it makes much difference whether the traveling twin moves at a constant speed through space for most of his journey; time dilation will happen anyway, even when there is acceleration or deceleration. It's only if time dilation were being exploited technically in space travel that it would be important to see what kind of accelerations we were dealing with: a complicated apparatus like the spacecraft certainly couldn't survive tremendous accelerations.

I suggest we look at an example. Let's assume that the spacecraft of the traveling twin leaves Earth with constant acceleration, and let it be the same as the acceleration of a free-falling rock on Earth. In the first second, the speed changes from zero to about 9.8 meters a second; in the second interval, it increases by the same amount; and so on. Keep this acceleration up, and you finally get to speeds great enough to make the effects of time dilation noticeable.

HALLER

A while ago, I set my students at the university a similar problem. You are correct, the effects of time dilation will quickly become noticeable. Let's assume the traveling twin starts as a young man in the direction of the Andromeda galaxy, which is about 2 million light years from Earth. Halfway there, he stops accelerating his spacecraft. By that time, 15 years have passed in the spacecraft.

He now decreases his speed at the same rate at which it had earlier been increased; in this fashion he'll arrive at zero speed in the region of Andromeda after 15 more years.

After his arrival, he decides to accelerate his craft back toward Earth. Finally, after a total of 30 years, he arrives. When he gets there, he notices that nobody remembers his twin brother. It turns out that 4 million years have elapsed on Earth.

EINSTEIN

Incidentally, the uniform acceleration and deceleration of the spacecraft would be useful for the space traveler in our case because he wouldn't have to deal with weightlessness. Since the acceleration would be the same as for a free-falling object on Earth, the space traveler would feel just the same as if he were on Earth. His acceleration would compensate in most, if not all, respects for the gravitational field on Earth.

NEWTON

In any case, Haller has made it clear that time dilation is important even for accelerations as small as those we experience on Earth—if only they take place for a sufficiently long period of time. I surmise that today's technology doesn't permit the construction of a rocket that can produce this acceleration over many years. What do you think?

HALLER

That's exactly the problem. We have no idea whether it will ever be possible. It will certainly be centuries before anyone is capable of using time dilation to demonstrate different aging processes in twins.

EINSTEIN (clearing his throat)

Gentlemen, my watch tells me it is after 6 P.M. The three of us are at rest with respect to each other. I can therefore assume we may neglect time dilation in our case, and your watches will show the same time. Shall we close the meeting for today and have some dinner?

His suggestion was unanimously accepted. Before long we were walking through the still crowded streets of Bern's Old Town in the direction of the Aarbergerhof.

TWELVE

Space Contraction

Next morning, the members of our small academy met at the usual time in Einstein's apartment. When I arrived, Newton and our host were already there. Einstein was smoking his first cigar, not a very good one, and pointed to the next room where Newton was preparing our morning tea.

EINSTEIN

Newton has just suggested a topic for today's session. Yesterday we spoke in great detail about time. Today it will probably be space. Newton believes he has found a contradiction concerning time dilation. I'll let him tell you the story himself.

At that moment Newton appeared, teapot in hand.

NEWTON

It's a good thing you're already here, Haller. We can finally begin. I would like us to look at a problem I encountered yesterday. It concerns time dilation, and I have already told Mr. Einstein about it.

HALLER

Why not? Solving a tough problem is better than listening to a lecture, even a lecture by Einstein.

I had spoken a little irreverently, knowing full well that Einstein agreed with me.

NEWTON

Let me explain my concern by returning to the muons and performing another thought experiment. Most of the muons that are generated by cosmic radiation in the upper atmosphere travel through space at a speed close to that of light. We know that time dilation is the reason they arrive on Earth at all, since their very short lifetimes would lead us to expect them to decay before arriving. Take, for example, a muon moving at a speed such that its gamma factor is 20. From the reference point of an observer at rest, the muon in motion lives 20 times longer than one that is at rest. Let the muon be produced at a height of 9 kilometers in a collision between a cosmic particle and an atomic nucleus, and let it then move vertically down toward the surface of the Earth. Thanks to time dilation, it will arrive without having decayed. To simplify, we'll assume that the muon decays the moment it reaches the surface of the Earth.

As he spoke these last words, Newton had cast a questioning glance at me, as though he weren't too sure about it. I responded.

HALLER

We have no trouble observing muons that decay at the precise moment they reach the surface of the Earth. True, only a tiny percentage of the muons decay right there. Many decay before reaching the Earth's surface; others penetrate into the ground before decaying. Some are slowed down by colliding with atomic nuclei and then decay more or less at rest. Anyway, I have no problem with your assumption.

NEWTON

Fine. I chose the height of 9 kilometers because a muon with a lifetime of exactly 1.5 microseconds and a gamma factor of 20 will move through space for exactly 9 kilometers before decaying. If you multiply 1.5 microseconds by 20 and by the speed of light (300,000 kilometers a second), you get 9 kilometers:

$$1.5 \times 10^{-6} \times 20 \times 300{,}000 = 9.$$

But here is my problem: Take an observer who is moving through space at the same speed as the muon, close to the speed of light. Let's assume Einstein himself is that observer.

EINSTEIN (smiling)

So be it. If it helps in the pursuit of truth, I'm prepared to move through space at almost the speed of light.

NEWTON

All right. Here you are moving along with the muon toward the Earth's surface; to be exact, the muon is at rest with respect to you, and the Earth is approaching a speed close to that of light—the speed defined by a gamma factor of 20. The muon that is at rest with respect to Einstein decays with a lifetime of 1.5 microsec-

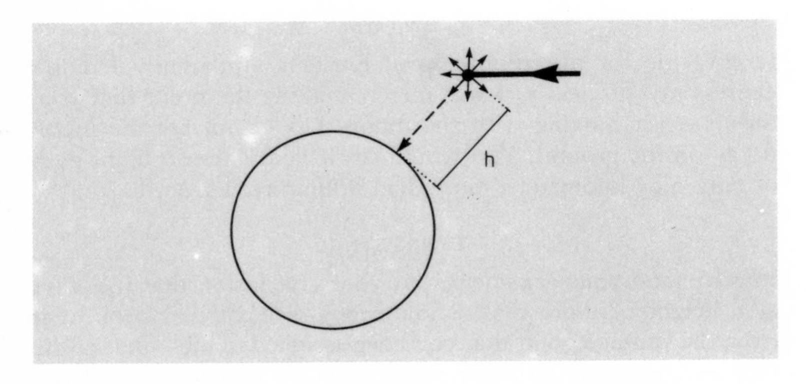

Fig. 12.1 A muon is produced at elevation h = 9 kilometers and moves vertically toward the Earth's surface with a speed corresponding to a gamma factor of 20 (i.e., almost at the speed of light). An observer on Earth measures a flight time of 30 microseconds before the muon hits the Earth's surface, where it decays. In its own reference system, the muon has a lifetime of 30 microseconds divided by the gamma factor of 20—that is, 1.5 microseconds. Thus an observer traveling with the muon would also see the muon decay upon impact onto the Earth, but would register a lifetime of 1.5 microseconds.

onds. During this time, it can't get very far; 1.5 microseconds multiplied by 300,000 kilometers a second yields exactly one-twentieth of 9 kilometers, or 0.45 kilometers. The muon wouldn't have a chance of reaching the surface of the Earth. There's a contradiction here: in the first case, the muon moves 9 kilometers through space; in the second, only 0.45 kilometers. So I must have made a mistake somewhere, since both distances can't be correct simultaneously. One of these numbers must be wrong.

EINSTEIN

My dear Newton, I can't quite agree that you have made a mistake. Let's look at your problem more closely. One thing is clear: The muon decays at a given point in space, say, exactly at its point of impact on Earth. This event is an objective fact; it can't depend on the observer. If it does decay just as it reaches the Earth's surface, the observer will notice it irrespective of whether he is sitting quietly on a chair next to that impact or racing through space at almost the speed of light. I therefore submit: the muon decays precisely upon its arrival on the Earth's surface, both for an observer at rest and for an observer moving with the muon.

NEWTON

Forgive me for interrupting you, Einstein. Apparently I didn't express myself clearly. I was merely making the point that you, the observer moving with the muon, would not see the muon decay on the ground. You would see it decay after a flight path of only 0.45 kilometers, more than 8 kilometers from Earth.

EINSTEIN

I understand your argument; but your conclusion that it decays at a height of more than 8 kilometers is incorrect. Look at it from the vantage point that you have assigned to me—that of the observer moving with the muon: Like the observer at rest, I will register the decay of the muon as it reaches the ground. We know that at that point the muon has moved only 0.45 kilometers. I agree with you there. But here is my decisive argument: time dilation means that time depends on the state of motion of the observer. Yet we haven't brought space into this argument, and now we must. I maintain that a change in the state of motion of the observer implies a change in the structure of space. More precisely, space will contract in the direction of motion; the rate of this change is the very same gamma factor that describes time dilation.

In our particular case, when I move with the muon through space, the point of the muon's creation appears to be not 9 kilometers from the Earth's surface but 9 kilometers divided by 20, that is, 0.45 kilometers. And this is precisely the distance the muon can cover at its speed of about 300,000 kilometers a second in its lifetime of 1.5 microseconds.

HALLER

Relativity theory claims not only the stretching of time by the appropriate gamma factor, it also means a shortening, a contraction, of space by that same factor. Only if both take place simultaneously can we be sure that the speed of light is universal in all reference systems. Only then can we be sure that events that happen at the same point in both space and time—like the muon's arrival on the Earth's surface and its decay there—will be described analogously in all reference systems.

While I was speaking, Newton had leaped up and run to the window. He gazed onto the busy street.

NEWTON

What on Earth have you done to space and time, Einstein? First you change the flow of time and make it depend on the observer; now you propose some equally degrading treatment to space. I have the feeling that practically nothing is left of my absolute space-time, the central theme of my *Principia*. Time as well as space is relative, depending on the observer—a dreadful thought.

HALLER

I can understand why you don't like some of Einstein's results, Sir Isaac. But it isn't really his fault. He has changed neither space nor time; he has simply discovered new aspects of space and time that we didn't know before. Time dilation and space contraction are facts that have since been demonstrated by experiment. They are direct consequences of the universality of the speed of light.

NEWTON

I realize that. After all, we're not a bunch of philosophers concerned with points of view and opinions, but natural scientists. The only thing that counts is experiment, and that's on your side, Einstein. Still, there is one thing I don't understand: distances in space are measured by yardsticks. Take this ruler, which is 30 centimeters long. How can its length depend on the state of motion of its observer? If I understand you correctly, this ruler is no longer 30 centimeters long but 30 divided by the gamma factor 20, or 1.5 centimeters, to the observer who is moving through space at the same rate as the muon we are looking at.

EINSTEIN

That's true, as long as the ruler is pointing in the direction in which I am moving. I assume that I'm descending toward the Earth's surface, just like the muon. Space contraction applies only in the direction in which the observer is moving. Let's just hope this experiment remains a thought experiment—otherwise I wouldn't be available to you after its conclusion.

NEWTON

When I was studying atomic physics a few days ago at Cambridge, I learned that the stability of matter, such as the matter that constitutes this ruler, ultimately depends on the stability of atoms. If I line up about a billion (10^9) atoms, I get a length of 10 centimeters. The reason this length doesn't change with time is simply the universality of the size of those atoms. It makes no difference whether we observe a hydrogen atom here on Earth or in a distant galaxy. It has the same structure and the same radius everywhere. This universality appears to me quite analogous to that of the speed of light. Einstein, as a fast-moving observer, establishes that the ruler is no longer 30 centimeters long but only 1.5, but I can't apprehend this from the viewpoint of an atomic physicist. The length of the ruler is determined by the number of atoms it consists of. To reach a length of 30 centimeters, I have to line up about 3 billion atoms. How do you expect me to contract the ruler, given that the number of atoms I have lined up can surely not depend on the observer's state of motion?

Einstein motioned to me, suggesting that I should respond.

HALLER

Of course the number of atoms doesn't change. Matter can't be created or annihilated just like that. Your problem, Sir Isaac, has a simple solution: the diameter of an atom—say, the 10^{-8} cm that we know to be the radius of a hydrogen atom under normal circumstances—is simply not a universal quantity, independent of the observer's state of motion. The contraction of space manifests itself on the atomic level as well. Atoms appear to be flattened in the direction in which they move.

Let's get back to our example. All the atoms in our rapidly moving ruler appear to be compressed in the direction they are moving in by the gamma factor, which we assumed to be 20. They are no longer spherical but ellipsoid, almost disk-shaped.

An unbiased observer might now ask what the shape of an atom really is. Is it spherical or disk-shaped? The answer must be both. The shape depends on the observer's state of motion. The structure of space and, hence, the shape of atoms and everything they may constitute are dependent on the observer.

NEWTON

You just said that the atoms that constitute our yardstick will contract. However can we establish that? The contraction of an object can only mean contraction with respect to a previously defined scale. If both the object and the scale contract, no effect can be perceived.

HALLER

I would agree with you if we always used the same scale to do our measuring. But in rapidly moving systems that's impossible. Let me remind you that we measure distances by means of the time it takes a light signal to travel that distance, multiplied by the speed of light. When we speak of the contraction of space, we are implying that all distances between points in space are measured by this means.

NEWTON

All right, Haller, I see your point. I had forgotten that distance is measured in terms of the travel time of light signals.

EINSTEIN

Haller, we've discussed numerous tests of time dilation, and it seems that no serious physicist today would dispute the effect. But what about space contraction? Has that effect been demonstrated experimentally? Not that I doubt the success of my theory! But you know that the only thing that really counts is experiment, no matter how elegant or brilliant the theory.

HALLER

Just like time dilation, space contraction can only be tested with systems that move very rapidly. And the only objects at our disposal that will do the job are fast nuclei or particles such as electrons or protons. We know that protons are extended objects, which we can view as little spheres; but the radius of a proton is minuscule in comparison to the radius of the simplest atom, that of hydrogen: it is ten thousand times smaller. Now let's make a proton moving at almost the speed of light collide with another proton, or with a nucleus. What happens in this collision is quite complex, and I won't go into it here. But we do know that specific details of this process will be different if the impinging proton looks like a sphere or a disk.

Some experiments of this kind were carried out at CERN. The results were unequivocal. Protons do behave like disks, and the faster they move, the flatter they get, just as your theory predicted.

Incidentally, the CERN accelerator wouldn't be able to function if it weren't for time dilation as well as space contraction. The effects of relativity theory were important elements in the construction of the accelerator. So relativity theory, at least as

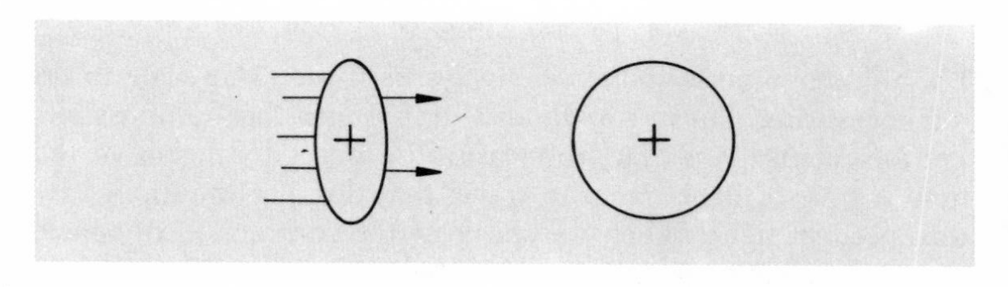

Fig. 12.2 A rapidly moving proton collides with a proton at rest. To an observer at rest with respect to the target proton, the incoming proton looks like a flattened ellipsoid: its mass distribution appears compressed in the direction of motion.

Fig. 12.3 A look at one sector of the tunnel that houses the SPS (Super Proton Synchrotron) accelerator at CERN. The proton beam runs in a vacuum pipe surrounded by magnets. They generate the magnetic fields that keep the beam running inside the ring-shaped beam tube. For the construction of such accelerators, the effects of relativity theory, especially time dilation and space contraction, have to be kept in mind. (Courtesy of CERN.)

far as the design and construction of particle accelerators are concerned, has become an engineering science.

Einstein appeared visibly relieved. He took out a bottle of white Neuchâtel wine and three glasses.

EINSTEIN

Please don't think I ever doubted relativity theory. But nothing is more satisfying for a physicist than the successful test of his ideas. Isn't it fantastic? In 1904 you spend countless hours at your desk, pondering space and time, and seventy-odd years later these same ideas are applied by engineers. They turn out to be essential for the construction of particle accelerators many kilometers long. I think we should toast that process! Here's to our science, which you, Mr. Newton, got started.

Although it was barely eleven o'clock, we decided to break up our meeting. We felt we deserved a walk along the Aare river. For days, Switzerland had been blessed by friendly summer skies, and we strolled along the river until we found an inviting restaurant for lunch.

THIRTEEN

The Marvel of Space-Time

After lunch we sat for a while without talking. Einstein stirred his tea absentmindedly, watching the tea leaves rally in the center of the cup. Finally, Newton spoke.

NEWTON

How strange that in the dimension of time the gamma factor signifies a stretching of the scale, with each time interval proportionately lengthened, but in space that same factor shortens the intervals—we divide rather than multiply by it. If I multiply the two factors by which I dilate time and contract space, I get the gamma factor divided by the gamma factor—in other words, one.

EINSTEIN (a little impatiently)

That isn't strange at all. It's simply a consequence of the universality of the speed of light. Otherwise, the speed of light would depend somehow or other on the observer.

NEWTON

That much is clear to me, Einstein. But I feel there is more to it. If the product of these two factors is always one, it could mean that neither space nor time remains unchanged when an observer modifies his state of motion, but that a third quantity does remain constant. You know, while writing the *Principia* I thought a lot about space. I was impressed by the fact that the definition of an object's position in space depends heavily on the coordinate system used, but the length of a distance does not. The length l of the distance between points A and B (see fig. 13.1), whose square is expressed mathematically by the equation

$$l^2 = (x_A - x_B)^2 + (y_A - y_B)^2 + (z_A - z_B)^2$$

$$[x_A\text{: } x\text{-coordinate of point A, etc.}],$$

is independent of the coordinate system. I can turn or move my coordinate system as I wish, and the length l, or rather its square,

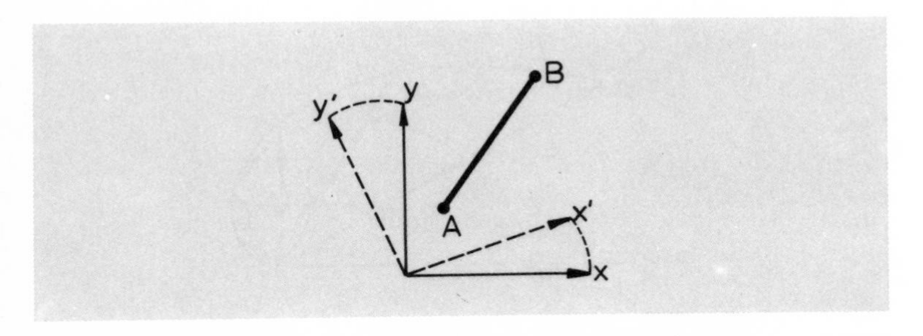

Fig. 13.1 The distance between two points A and B is independent of the coordinate system in which it is described. It can be determined in terms of the *x*, *y* coordinates, or in terms of the *x'*, *y'* coordinates, also shown in the figure, which are rotated with respect to the *x*, *y* set.

will not change; it is invariant. In the same fashion, the length of a distance doesn't depend on the observer's state of motion.

But in relativity theory the whole picture changes. We have learned that the length l between two points in space is not an absolute quantity but is dependent on the observer's state of motion. The same goes for the time difference between two events. Nevertheless, I think your theory contains some specific quantity that will indeed remain unchanged—even though the position of the observer, or of the coordinate system, does change.

EINSTEIN (casting an appreciative glance at Newton)

Sir Isaac, once again you're on the right track. There is certainly something in relativity theory that doesn't change when you shift to another frame of reference, something that is the same for all observers. Let me start with another little thought experiment. Let's return to those hypothetical one-second particles with which we began our explanation of time dilation. I'll assume I have particles that I can shoot into space from some space station, and that I can choose any speed I please, with the sole limitation that it be smaller than the speed of light.

The particles decay exactly one second after being launched, which means they have a lifetime of one second. This measure of their lifetime is correct only in the reference system of the particle itself—that is, in the coordinate system that moves along with it. In the rest system of the spacecraft, on the other hand, time dilation makes the particles that have been shot off live longer. In that system their lifetime is one second multiplied by the gamma factor.

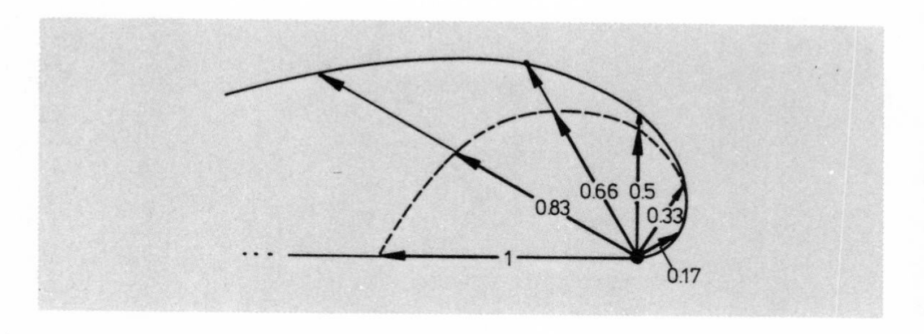

Fig. 13.2 A spacecraft emits particles of a one-second lifetime in various directions. The speeds of these particles increase with the direction in which they are emitted from right to left in our example; they are marked, in units of the speed of light, in connection with each arrow. (The arrow marked 0.5, for example, stands for a particle traveling at 150,000 kilometers a second.) Without time dilation, the distances between emission and decay would be the products of velocity and lifetime; they are shown as heavy arrows, the envelope of which is the broken spiral. If the particles were emitted at precisely the speed of light, their path length would be exactly one light second (see the arrow marked 1). Time dilation increases the paths from emission to decay by the gamma factor. This is indicated by the lengthened arrows, with the full spiral as their envelope. Here the limit of the speed of light cannot be reached—it would lead to an infinitely long path. (Note that the lifetime of the hypothetical particles is used here as a classical lifetime; that is, the decay takes place right after the time has passed. For purposes of simplicity, we have ignored the fact that this does not accord with quantum theory, where only a probability of the decay can be given.)

Einstein took a sheet of paper and started to sketch the different trajectories of particles emitted from the spacecraft (see fig. 13.2). He explained his sketch, continuing his lecture.

EINSTEIN

Clearly, the length of the particle trajectories will depend on their initial speeds. If that were zero, a particle would just sit at the origin and decay there after one second—it wouldn't travel at all. Now it's interesting to consider various speeds in a space-time diagram where, for simplicity's sake, we'll take space to have only one dimension; we'll call that the x axis and ignore the other spatial dimensions.

Einstein took another sheet of paper and drew a space-time coordinate system (see fig. 13.3).

EINSTEIN

In this system I'll now enter the only two events that are important for our one-second particle—its creation and its decay. Since the

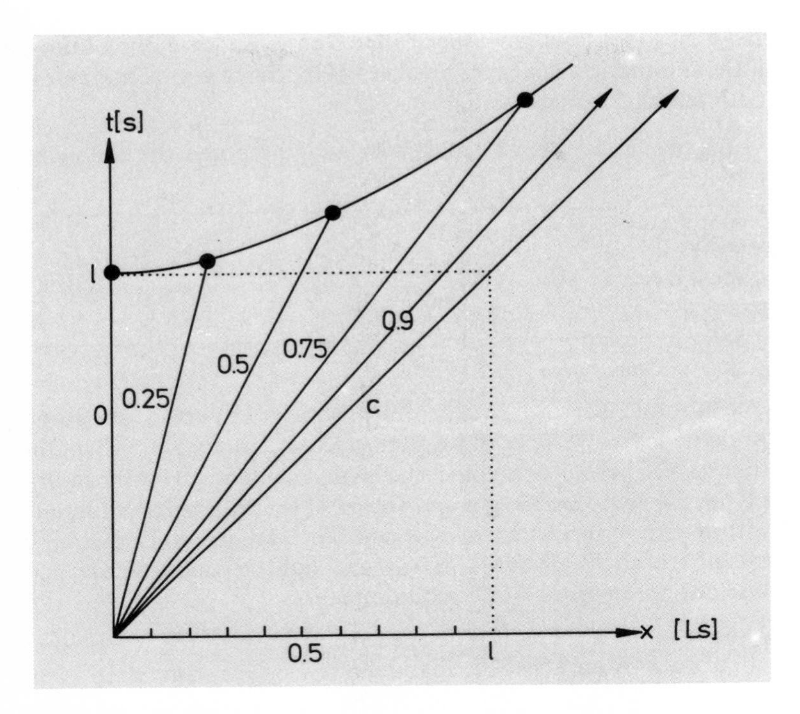

Fig. 13.3 The space-time coordinate system into which Newton entered the world line of several particles with one-second lifetimes. For a particle at rest, the world line runs along the time axis from its production event (in the origin) to its decay event at $x = 0$, $t = 1$ second. In the absence of time dilation, all decay points would appear on the dotted line at $t = 1$ second, for particles of various states of motion. Time dilation makes them appear along the curve, or hyperbola, instead. The individual world lines correspond to the cases in the table.

particles, during their short life span, move uniformly through space along straight lines, the corresponding world lines will be straight lines in our space-time system.

Newton took the pencil from Einstein's hand and began to draw a few world lines into the sketch.

NEWTON

Let's assume I interpret the origin of your space-time system as the event of the particle's creation. Suppose the speed of the created particle is zero, so that the particle remains at space point $x = 0$; its world line will simply be a straight line starting at the origin and moving along the time axis to point $t = 1$ second.

Now let's look at a few other cases. Don't you have one of those little automatic calculators, Haller? Why don't you compute the path length for a few initial speeds?

I took out my pocket calculator and computed the following table:

speed v	0.25	0.50	0.75	0.9
gamma factor γ	1.03	1.16	1.51	2.29
path length x	0.26	0.58	1.13	2.07

This table gives the speeds of the particles in units of the speed of light, the corresponding gamma factors, and the path lengths they travel before decaying. The path length is given by multiplying the speed by the gamma factor, since the particle's lifetime is supposed to be exactly one second. The path length is measured in units of light seconds. Thus, 0.26 light seconds = 0.26 × 300,000 kilometers = 78,000 kilometers.

Newton now added to the space-time diagram the world lines of the various particles. He marked a few points, connected them with a hand-drawn curve, and stared at the diagram in fascination.

NEWTON

Since the lifetime of these particles is exactly one second in their own rest system, their actual lifetime in the observer's system is given by the gamma factor. That means $t = \gamma$. Professor Haller, would you kindly compute the quantity $(t^2 - x^2)$ for all of the values in your table?

I knew, of course, what Newton was driving at. I complied with his request and computed:

$$(1.032)^2 - (0.262)^2 = 0.996$$
$$(1.162)^2 - (0.582)^2 = 1.012$$
$$(1.512)^2 - (1.133)^2 = 1.002$$

EINSTEIN (breaking in)

I don't think Haller need go on computing, Newton. The result is clear. Your difference between the squares of the time and space coordinates of the event of the decay, $t^2 - x^2$, is a constant; and in our case this constant is equal to one. The fact that Haller's calculator did not produce the value of exactly one is simply due to rounding errors.

NEWTON

I realized this regularity when I started drawing the curve in the space-time graph. I've seen plenty of curves like that in my studies of planetary motion. But let me calculate that difference exactly. It's

$$t^2 - x^2 = \gamma^2 - (v\gamma)^2 = \gamma^2(1 - v^2) = 1,$$

because the square of the gamma factor γ is merely $1/(1 - v^2)$, where v is the speed of the particle in question in units of the speed of light; v is therefore a pure number.

Newton beamed at us; he was clearly relieved. He then continued.

NEWTON

Einstein, you really should have told me right from the start how simple matters get when we look at events in space-time. I think we should give your theory of relativity a new name: the theory of the absolute. What counts in physics is that which is absolutely valid, that which is not dependent on the position of an observer. And that's just what we have found—the square of the time coordinate minus the square of the space coordinate. This number is the same for all observers. It is absolute; it is an invariant of space-time. It's no wonder we have the strange phenomena of time dilation and space contraction. Everything serves one purpose only, to keep this difference constant at all costs. Relativity be damned! Long live the absolute! Forget about space, forget about time—from now on let's only refer to the two of them in tandem, space-time, let's focus only on the quantity that is independent of the observer, the difference between the squares of time and space.

Neither Einstein nor I had ever seen Newton so animated. He had clearly been converted into a true disciple of the theory of relativity.

EINSTEIN

I have no objection to your renaming relativity theory. I didn't like the name myself at first. But I'm afraid it's too late to change the name, so we'd better leave it as is. But I agree that from now on we should speak of space and time in one and the same way—as space-time. This suggestion, by the way, is not really mine. It was made by Hermann Minkowski, my former math professor in Zurich, when he was teaching at Göttingen University in Germany, three years after I published my paper.

Now back to our problem. For the $t^2 - x^2$ difference, it would be more accurate to write $(ct)^2 - x^2$, explicitly mentioning the speed of light c, whenever we're measuring space not in terms of light seconds but in the usual units such as meters or kilometers. And we can now take all three coordinates of space into account. Then the difference we discussed will look like this:

$$(ct)^2 - (x^2 + y^2 + z^2).$$

We can now see what really happens when we move from one reference frame to another one which is in motion with respect to the first. We already know that the flow of time as well as the length of any distance changes in this case. Space and time can be thought of as somehow being rotated into each other. The quantity that doesn't change in this process is the difference between the squares of space and time.

That reminds us of the rotation of a spatial coordinate system: the rotation will change the coordinates of individual points but will leave the distance between any two of them invariant. You've already pointed that out to us, Newton.

If we replace the length of a path, or, more precisely, its square, by the square of time minus the square of the length, we find ourselves firmly in relativity theory. The transition from a reference system at rest to a system in rapid motion can therefore be seen as a quasi rotation of space-time in such a way that the difference as we defined it remains unchanged.

It isn't all that surprising that the difference between the squares of time and space of two events is the only thing that remains absolute and doesn't change with the state of the observer. Let's look at another space-time diagram.

Einstein produced and commented on a new diagram (see fig. 13.4). He spoke of events that he called lightlike, timelike, or spacelike relative to one another in space-time.

EINSTEIN

You can see that the difference between the squares of time and space vanishes for all events that can be reached from the origin through a light signal; we call these events lightlike with respect

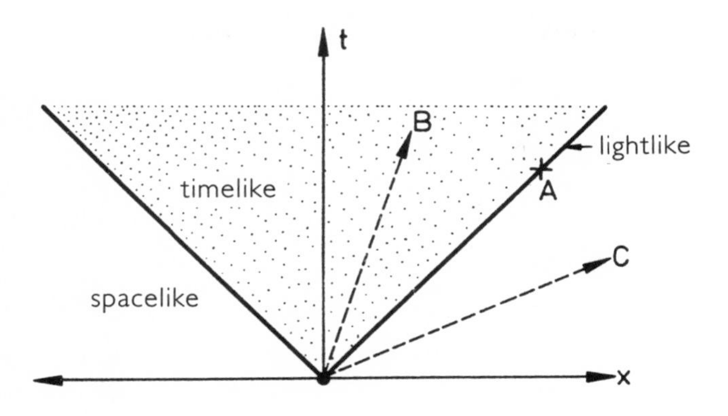

Fig. 13.4 Space-time diagram with one space axis. Every point in the *t-x* plane stands for an event. The totality of events can be classed in three groups by comparison with a light signal emitted from the origin: its world line consists of the two straight lines issuing from the origin at 45 degrees. This angle results from the fact that we measure distance in space in units of light seconds.

The two straight lines denoting the light signal are called the light cone; they contain all events for which the difference $(ct)^2 - x^2$ vanishes. That is why they are also called lightlike events, like the event here denoted A. Events inside the light cone (in dotted area) will correspond to positive values of the above difference, such as point B or any point along the time axis itself. These events are called timelike.

The remaining events, which are located outside the light cone, like point C, are distinguished by a negative value for the difference between the squares of temporal and spatial coordinates. They include events on the space axis itself and are called spacelike.

to each other. If I send a laser beam from Earth to the moon, the two events defined by the emission of the signal and the arrival of the signal are lightlike with respect to one another. This is not dependent on the state of the observer; we know that the difference between the squares of time and space doesn't change when we shift to another reference system. An observer watching the path of our laser signal from Earth to the moon while located in a spacecraft that is passing by the Earth in rapid motion will observe these two events as lightlike with respect to each other, just as we do while sitting on Earth. This is a direct consequence of the universality of the speed of light.

NEWTON

I understand. If there were no such universality, it would make no sense to classify the events relative to one another as lightlike, timelike, or spacelike.

EINSTEIN

Let me give you another example: let's consider an event here on Earth, such as the birth of Christ in the year 0, and another event, such as the eruption of a volcano on the planet of the star Vega (26 light years away from Earth) in the year A.D. 30. The difference $(ct)^2 - l^2$ is rendered by $30^2 - 26^2 = 15^2$; it is timelike. This difference is independent of the reference system. If an astronaut had registered these two events from a rapidly moving spacecraft, the difference would have been the same in his reference system.

Let an astronaut be passing the Earth in his spacecraft in the direction of Vega exactly at the time of the birth of Christ, and let him arrive at Vega exactly at the time of the eruption of the volcano. Since that distance is 26 light years, the astronaut has to travel at almost the speed of light in order to get there on time. We can immediately tell how many years it will take the astronaut in his spacecraft to get from Earth to the planet Vega. In his system, the spatial distance between the two events is zero, because at the time of the birth of Christ he is close by the Earth; and when the volcano erupts, he is in the region of Vega. The time distance between the two events is, of course, not zero; we determined above that it is about 15 years. That means the astronaut will have aged by 15 years by the time he zooms past Vega.

NEWTON (nodding approvingly)

It makes me happy, Mr. Einstein, that even in your theory not everything is relative. Now that we have seen an absolute quantity in it, the difference between the squares of time and of space, I'm convinced that your theory is correct. But that quantity is really an odd one. Whoever would have thought, in my time, that this difference would play a central role in physics? Certainly not I, certainly not Leibniz, much less anybody else.

HALLER

The difference between the squares of time and space not only implies a union of space and time; it also indicates an important difference between the two. We are dealing with a difference, not with a sum. Had we defined the sum of the squares as the invariant, we could speak of a true unification of space and time. But that is not the case here. Although a change in the observer's state of motion will mix up space and time, there remains a basic difference between space and time in all reference systems, and we have seen how that leads to the phenomena of time dilation and space contraction. This difference manifests itself in the minus sign between the squares. We were hinting at that when we said, a while ago, that space-time doesn't have four coordinate axes but three plus one—three for space, one for time.

NEWTON

What an odd structure. Why is it so, Einstein? Why is the inner structure of space-time defined by this strange difference? How about you, Haller? Does it make sense to you?

HALLER

You're asking a lot of me, Sir Isaac. To this day, nobody knows why there are three dimensions for space but only one for time. Nor does anybody understand why space and time are welded together by the theory of relativity. The only thing that's certain is that we can derive the basic properties of space and time from simple facts such as the universal speed of light. And that's all.

Even today, science is far from providing answers to these questions. Sometimes it seems miraculous to me that we can even ask the question about the basic structure of time and space. Our world appears to be put together in a much simpler way than we might have suspected from our daily experiences. The blueprints of creation seem somehow to be shining through the fundamental structure of space and time. They show a certain simplicity and a symmetry—although these qualities are by no means easy to decipher. John Wheeler, one of my colleagues in theoretical physics, once said that if ever we manage to determine the laws of the universe, including those that govern space and time, we'll be surprised that they weren't self-evident from the start; it's so hard to discover simple things.

EINSTEIN (having lit a cigar)

Newton, I always thought you were a talented pragmatist. As far as I know, you never asked where that odd law of the universal attraction of masses, which you discovered, hails from—you certainly didn't raise the question in your book. Remember your statement *Hypotheses non fingo?*

I suggest we stick with your principle and don't attempt to explain the ultimate origin of the structure of space and time. Let's accept it as a given, and concentrate instead on identifying its consequences. There are many that we haven't discussed yet.

Among them are phenomena connected with the dynamic properties of matter; in my opinion, this is where we find some of the most interesting and instructive implications of the theory of relativity.

HALLER

I couldn't agree with you more. In our discussion about relativity theory we've reached the point at which we have to begin talking about matter. From the start I had planned to hold some of our discussions at the CERN research center, so this might be a good time to break up our academy sessions here in Bern and move to Geneva. I've already made some arrangements. If you don't mind, I'd like to work out the details of our trip.

It was already late in the afternoon. We agreed to leave for Geneva next morning, which happened to be a Sunday, and to go in my car. I had made reservations for three rooms in the CERN hostel.

FOURTEEN

Mass in Space and Time

The warm, sunny weather that Switzerland had enjoyed for several days remained with us, and we got off to an early start on Sunday. We reached the steep embankments of Lac Léman, the Lake of Geneva, not far from the town of Vevey. I parked briefly so we could take in the panoramic view of the Alps on the other side. Deep below us, the lake shimmered in the morning sun. This stopping place offers one of the finest views in Europe.

Newton had appeared somewhat bored on the trip, but the view into the Rhone valley, toward Martigny, captivated him. Finally he turned toward Einstein.

NEWTON

Yesterday, you and Haller hinted that relativity theory would have a few more surprises for me, especially with regard to the dynamics of matter. Last night I tried to make some sense out of your hints, but I'm afraid I didn't get very far. Since you've thoroughly straightened out my definition of space and time, I imagine certain other concepts of my mechanics may also need to be revised rather radically. Let's begin with the concept of mass. I wonder what relativity theory would put in the place of an object's mass, normally measured in grams or kilograms.

EINSTEIN (smiling)

My dear Sir Isaac, I understand only too well that you are feeling unsure about the concepts and definitions that play a role in your theory of mechanics and dynamics. At this point I don't want to go into detail; but I assure you that talking about the mass of an object makes sense in the framework of relativity theory too. In fact, with one important modification your ideas of mass remain essentially valid in relativity theory.

As Newton and Einstein were speaking, I had resumed driving. They continued to discuss various aspects of relativity theory. We soon left the freeway, skirted Geneva Airport, and found ourselves on the Meyrin Road, which connects the city of Geneva with the CERN laboratories. When we reached the main gate of this large complex, I parked my car and picked up the keys for our rooms in the guest hostel. A few minutes later we passed the security guards on our way into the compound.

EINSTEIN (looking out of the car window)

This is supposed to be a *physics* institute? In my time, all the apparatus for experimental physics could easily fit into a few rooms. This place looks more like a factory than a research lab. Can anyone actually do research in such a gigantic place?

To me, research means above all the freedom to think and work without interference. I can hardly imagine that this enterprise can function without an overwhelming bureaucratic organization; at the very least, it must require long-range planning. But research, the way I understand it, can't be planned over extended periods. You need ideas, imagination—and everyone knows that these are commodities that can't be planned. If available at all, they occur spontaneously, and usually not at the most opportune time.

HALLER

There is no doubt that research here at CERN can only be carried out by planning individual experiments long in advance; that's the trend of our times. Besides, we can only keep the costs of research in check by careful planning—and that's certainly needed. Research institutes like CERN, after all, are supported by tax money. Experiments in physics can no longer be done the way they were in the nineteenth century, in Faraday's time. They last months, sometimes years, and physicists usually work on a single project, which isn't necessarily a bad thing. Many scientists consider it advantageous to work in collaborations made up of smaller teams from universities in different countries. On the other hand, Mr. Einstein, although like most theoreticians you don't enjoy working in a large group, you would have no problem here. If you were to work at CERN, no bureaucratic constraints would be imposed on you. Even at a research institution as large as this one you would have complete freedom to follow your own interests. You probably had less freedom at your Bern patent office.

EINSTEIN

Consider yourself lucky, my dear Haller, that my boss at the patent office didn't hear that. He too was called Haller, incidentally, Friedrich Haller, and I had no complaints about him. He gave me all the freedom I could have asked for. I certainly had more leisure to pursue my research interests than I would have had at a university, where every young scientist is under pressure to publish a lot of papers in a short time—"publish or perish." Academics today hardly have time to read their colleagues' papers. And you know what kind of work that produces—mostly stuff that should go straight into the wastebasket.

Our conversation was interrupted by Newton, who pointed to a street sign we were just passing: Einstein Road. Einstein was not impressed.

EINSTEIN

I hope you're not jealous, dear Newton. Obviously my theory was of help here. Still, I bet there's a Newton Road here too! Isn't that right, Haller?

Instead of answering I slowed down a little and pointed to the street sign on our right: Newton Road. That clearly satisfied Newton. Still, I had the impression he was quietly sizing up "his" street, a little side road leading up to a long lab building, and that he was comparing it with the longer and wider Einstein Road. In a few moments we reached the hostel, and our trip came to an end.

We quickly settled in. It was Sunday, and we decided to use the free time for a trip to the nearby Jura mountains. After an hour's drive from CERN, passing through the little French town of Gex and winding our way up the road to the Col de la Faucille, we reached the high plateau of the Juras. We took a hike through the Alpine pastures and soon found ourselves at the rim of the Jura mountains that falls steeply toward the Geneva basin. From this vantage point, we had a spectacular view of the city of Geneva, the lake, and the French Alps beyond it. We sat down and continued the discussion we had started in the car.

NEWTON

Now it's time to do some explaining, Einstein. How do we handle mass in relativity theory? It must be quite difficult to incorporate the concept of mass into your theory. I well remember how I defined the mass of an object in my day. Given that matter consists of atoms, and that atoms—according to our present view—consist of nuclei and electrons, I suppose we could think of the mass of a macroscopic object simply as the sum of the masses of its nuclei and electrons. So we can restrict ourselves to thinking about the masses of electrons and atomic nuclei.

Now let's assume I accelerate a proton—the nucleus of the hydrogen atom—in the CERN accelerator down there in the valley. We already know that the relativity of space and time doesn't permit us to accelerate a proton to a speed greater than that of light. I'm convinced there must also be a dynamic reason why this can't be done. But I just don't see what the reason can be if there's no change in the concept of a particle's mass in relativity theory. In principle it should be possible to accelerate the proton to ever higher speeds. If the process is continued long enough, the speed of light should be reached or exceeded at a certain point. On the other hand, that can't be done because the speed of light must not be exceeded. So there is something wrong here. I have the feeling that the mass of the particle may change at very high speeds; more precisely, I think the mass might increase in such a way that it becomes impossible, even in principle, to accelerate particles to speeds beyond the speed of light.

EINSTEIN

You have a knack for quickly getting to the heart of the matter. Remember, I told you this morning that your concept of mass can quite simply be taken over to the theory of relativity, though with a slight change. That change is exactly the effect you just suspected. At very high speeds, the mass of a particle does increase, and the theory says exactly how that happens.

HALLER

If I may, I will describe the effect by a little thought experiment. Since I have to introduce observers moving almost at the speed of light, we had better move out into space; but we'll have to take a rifle and a wooden plank with us.

I now made a drawing (see fig. 14.1).

HALLER

Suppose we fix our location at a distance of one kilometer from this plank, which is suspended in space, and suppose we are at rest with respect to the plank. We now fire a bullet toward the center of the plank. Let the bullet move through space at a rate of a thousand kilometers a second. Exactly one second after being fired, the bullet hits the plank, penetrates it a certain amount, and

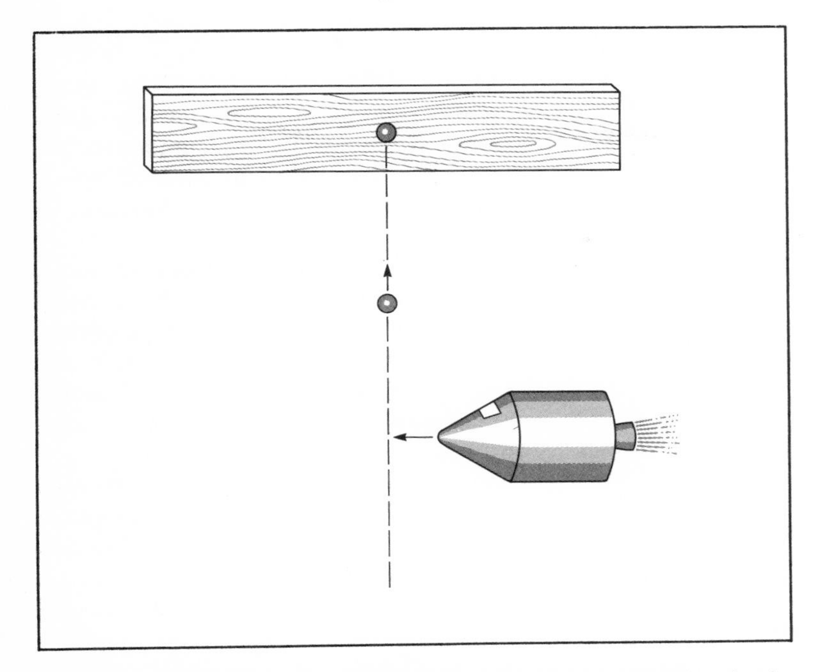

Fig. 14.1 A bullet's trajectory before it hits a plank. Its penetration depth depends on the bullet's momentum, that is, the product of its mass and its speed. The greater the bullet's speed, the deeper the pit. At constant speed, a heavier mass will penetrate more deeply, so a lead bullet will make a deeper impact than a steel one of the same diameter because lead is heavier than steel.

To an observer registering these happenings from a rapidly passing spacecraft, the bullet would appear to be moving more slowly; this is a consequence of time dilation.

sticks in the wood. Now the plank moves away from us, since the bullet has transferred its momentum to it. The depth of the penetration depends both on the speed of the bullet and on its mass. The greater its speed, the deeper the bullet penetrates. At constant speed, the depth of the penetration increases with the mass. A steel bullet won't penetrate as deeply as a lead bullet of the same size, because lead is heavier than steel.

NEWTON

Why not just say that the depth of the penetration depends on the momentum of the bullet, that is, on the product of mass and speed?

HALLER

That's correct, it is the momentum of the bullet that matters here. But now let's look at it from the vantage point of a passing spacecraft. Let this spacecraft be moving very rapidly, almost at the speed of light; to be precise, let's give it a gamma factor of 10. And let it be moving parallel to the plank, that is, perpendicular to the direction of the bullet.

We know that space contraction, as it follows from relativity theory, does not affect the direction perpendicular to that of motion. The observer in the spacecraft will see, just as we do, that the plank is exactly one kilometer away from the rifle. The only difference is that to this observer neither gun nor plank is at rest—they are both zooming past the spacecraft at almost the speed of light.

And here comes the essential point: because of time dilation, the observer in the spacecraft will observe that the bullet doesn't race through space for one second but for one second multiplied by the gamma factor, which makes ten seconds. From his viewpoint, the bullet moves toward the plank not at one kilometer a second but at only a hundred meters a second.

NEWTON

Just a moment. We, the observers at rest with respect to the plank, see the bullet penetrating the wood; and if we haven't seen it happening, we can easily check after the fact. Suppose I fired the bullet at the plank not at a thousand meters a second but rather at the relatively modest speed of a hundred. In that case the bullet would penetrate only a short way into the plank. But how deeply the bullet penetrates the wood is an objective fact, which can't depend on the observer, since the wood has manifestly been damaged in our experiment. The observer in the spacecraft might be astonished to see that the bullet, despite its relatively slow motion, causes so much damage. Pretty strange, isn't it?

EINSTEIN

Whether the observer on board the spacecraft is surprised or not depends on his knowledge about relativity. If he is a convinced disciple of Newtonian mechanics, he will certainly be surprised. But if he believes in my theory, he won't be surprised at all. The reason is obvious. You rightly said that the damage done to the wood—or rather the penetration depth of the bullet into the plank—can't possibly depend on the observer. But as we said before, the depth depends on the momentum of the bullet, which is the product of its mass and its speed. A slower bullet can very well cause the same damage as a faster one if it is heavier. Having said this, we have hinted at the solution of the problem. The quantity that is important here, which must not under any circumstance depend on the observer, is the momentum—the product of mass and speed. Since the speed decreases by the appropriate gamma factor, the mass has to increase by that same factor.

At rest, the bullet has a mass that I'll call m. Incidentally, this is exactly the mass the bullet would have if Newtonian mechanics were fully valid.

NEWTON

Since we're dealing here with low speeds compared with the speed of light, I conclude that your mass m is identical with what I considered to be the mass of a body when I wrote the *Principia*.

EINSTEIN

Precisely. To make sure we know what we're talking about, we call this mass the rest mass of a body. But we could equally well have called it the Newtonian mass. Now, if that body moves very fast, its mass—or, more accurately, its moving mass, which I'll call M—increases by comparison with the rest mass m, and the increase corresponds to the gamma factor:

$$M = \gamma m = \frac{m}{\sqrt{1 - \left(\frac{v}{c}\right)^2}}$$

The mass increases in the same way as a time interval, which is stretched by the same gamma factor.

To illustrate this effect, I sketched the ratio of the moving mass M and the rest mass m (see fig. 14.2).

HALLER

The increase in mass, which is often called the relativistic mass increase, accelerates as the body approaches the speed of light. But it's impossible to accelerate the body all the way to the speed of light since its mass would then increase to infinity. To bring that about we would have to apply infinite energy, and that is beyond anyone's means.

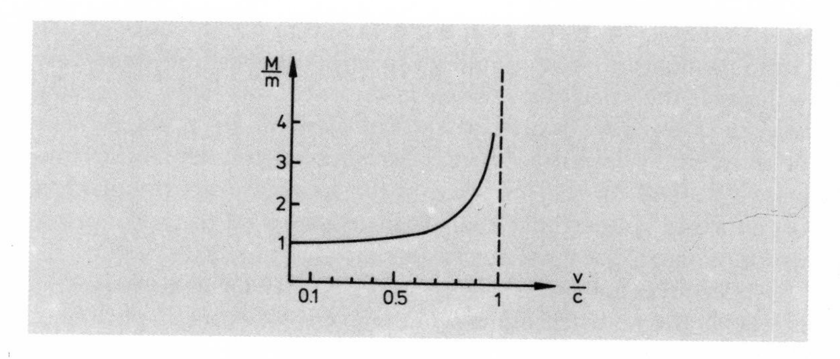

Fig. 14.2 The mass M of a moving object increases from its value at rest. This effect is shown here as a function of the object's speed, in units of c, the speed of light. Like time dilation, it becomes noticeable only when the speed becomes comparable to that of light. The closer the mass gets to the speed of light, the more rapidly it increases. The extreme case $v = c$ will never be reached, because that would entail the mass's becoming infinite.

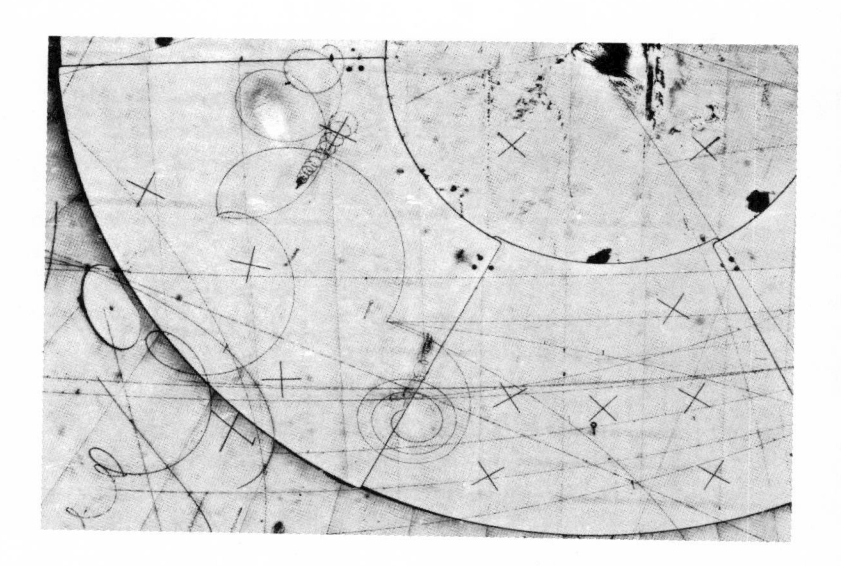

Fig. 14.3 The particle tracks in a bubble chamber are curved because of the action of an external magnetic field. The curvature depends not only on the speed of the particle but also on its mass—or, rather, its moving mass. Thus the relativistic increase in mass can be observed directly. (Courtesy of CERN.)

EINSTEIN

You can see, Newton, that with relativity theory things happen exactly as you guessed earlier. The mass increases, and it is impossible, even in principle, to accelerate a massive object all the way to, or even past, the speed of light.

HALLER

The large accelerator at CERN, which you can see down there in the Geneva basin, accelerates protons pretty close to the speed of light. Their final energy is then virtually independent of their speed; it's dependent only on their moving mass. If you run the accelerator at half-power, the particles are accelerated to an energy just one-half of what they reach if you run the machine at full power. In both cases the speed of the particles is very close to that of light, only their mass is greater when the machine is run at full power, and that's why they have greater energy. In atomic physics and in particle physics, we usually specify the energy of a particle in terms of electron volts, abbreviated eV.

NEWTON

I know that. An electron volt is the energy that one electron acquires as it goes through an electrical field with an applied voltage of 1 volt.

HALLER

The energy you impart to a proton by means of the CERN accelerator when it's running at full power is 400 giga electron volts—usually abbreviated GeV. That amounts to 400×10^9 eV, which is quite a lot. The speed of the particles is then very close to the speed of light c, more precisely $0.9999973\,c$. We can now calculate the mass M of the moving particles in the accelerator:

$$M = \gamma m = \frac{m}{\sqrt{1 - (0.9999973)^2}} = 430m\,.$$

NEWTON

My goodness! Those protons in the machine down there actually move around with a mass about 400 times greater than their rest mass m. Is it possible to observe this enormous mass increase directly? Surely the effect should manifest itself somehow.

HALLER

And so it does. The protons move in a large, ring-shaped tunnel. Keeping them on course requires magnetic fields; without an external force a proton would move in a straight line through space, like any other particle.

NEWTON

I see what you're getting at. The strength of the magnetic field determines the extent to which the particle flying through it will change its direction. But the particle's mass is also essential: the larger the mass, the less the influence of a given field. As soon as the relativistic mass increase kicks in—that is, as soon as the protons approach the speed of light—much stronger magnets will be needed to keep them on course.

HALLER

Right. If there were no mass increase, we wouldn't need such strong magnetic fields at CERN; the protons could be kept on track with fairly weak magnets. But because of the mass increase, the magnets need to be 430 times stronger than they would be without that increase. The huge magnetic fields that are needed can be reached only by feeding in the required energy in the form of electric energy. The energy CERN consumes for this purpose corresponds to the entire output of a medium-sized power plant. They could run their accelerator much more cheaply if the effects predicted by relativity theory didn't actually occur.

EINSTEIN (smiling)

I'm really sorry that my theory raises the price of the accelerator. Your theory, Newton, would save a lot of money if it were applicable here. I hope I won't run into the director while I'm visiting CERN. He might ask me to reimburse him for all these additional expenses.

HALLER

I think we can do without that encounter. But I would like to mention an interesting effect here. Particle detectors such as cloud chambers or bubble chambers make the particles' tracks visible. In the presence of a magnetic field, a particle's track is curved, and the degree of curvature is dependent on the mass—or, rather, on the moving mass M—of the particle. In this way the relativistic mass increase can be observed experimentally.

Newton was lying in the grass with his head resting on his elbows, surveying the Geneva scenery from above.

NEWTON

Isn't it odd? Down there they are using vast amounts of energy to accelerate normal matter—in other words, protons—almost to the speed of light. But we have other particles, namely photons, light particles, that by definition move through space at the speed of light. They have energy but no mass. In terms of rest mass, which you defined earlier, Einstein, photons are particles without a rest mass. They only have energy. Putting aside the fact that protons have rest mass but photons don't, I believe those two kinds of particles have something in common: both are subject to the limitation imposed by the speed of light.

EINSTEIN

Why do you think that's odd? We know, after all, that the speed of light is not just the speed of photons but a kind of fundamental speed in our universe.

NEWTON

I realize that. But as I see it, protons moving almost at the speed of light—like the ones down there at CERN—look almost like photons. There doesn't seem to be much difference between a beam of rapidly moving protons and a photon beam of about the same energy.

Please don't get me wrong—I'm just baffled at the concept of mass. When I wrote the *Principia,* I assumed that I knew exactly what mass was. Now, after all that I've learned from you about the structure of space and time, I'm afraid I've lost all feeling for mass. Wouldn't it be easier if all particles, including protons, had no mass, like photons? Why do massive particles exist at all?

I'm also puzzled about the strange connection that appears to exist between energy and mass. The kinetic energy of an object depends, as we know, on its mass and the square of its speed—$E = \frac{1}{2} mv^2$. This equation applies as long as the speed of the object is much smaller than the speed of light c. If you raise the speed of a bullet by a factor of 2, the kinetic energy it carries increases by a factor of 2^2, that is, by a factor of 4. But let's get back to those protons moving around at CERN almost at the speed of light. What is their energy? In other words: How should my equation $E = \frac{1}{2} mv^2$ be adjusted in a case where v is close to the speed of light c?

If I increase the energy of the CERN protons, we already know that I increase only their moving mass M, since their speed barely changes. Energy and mass are therefore directly proportional to each other. That's really curious—it might make you suspect a secret connection between mass and energy, a mass-energy relation. Why are you so quiet, Einstein? What's your opinion?

EINSTEIN

I have a quite definite opinion. The relation between energy and mass that you just mentioned is a very real one. It plays a specific, major role in relativity theory, and I have a lot to say about that. Without exaggerating, I believe we can say that this relation may well be the most interesting facet of the theory of relativity.

But we've gone well past noon, gentlemen. Shouldn't we start our picnic? For my part, I'm ravenous.

And so, for the time being, we abandoned the search for that formula—the formula that was to describe the possible relation between mass and energy in relativity theory. Instead, we munched on the sandwiches I had brought from the CERN cafeteria.

FIFTEEN

An Equation That Changed the World

After the picnic we walked for a while along the crest of the Jura mountains. Newton was walking in silence, while I was praising the merits of the Jura range to Einstein. Few places in Western Europe can offer the solitude and the untouched nature of the Jura forests. When I got stuck on problems in my work at CERN, I would frequently come up here for fresh ideas. We now sat down by a wooded patch, and Sir Isaac immediately confronted us with a problem he must have been pondering for some time.

NEWTON

The connection between mass and energy is buzzing around in my head. Now here is my train of thought: the energy of a rapidly moving object—an object moving almost at the speed of light like a proton in the CERN accelerator—is likely to be proportional to the moving mass M. If I double its mass, its energy will double also. Now we know that the moving mass M increases steadily as we gradually approach the speed of light, which implies that the energy of the object also keeps increasing. On the other hand, the equation we're looking for should contain the square of the speed, just like my old formula $E = \frac{1}{2} mv^2$; for the energy that is due to the motion of an object is, for dimensional reasons alone, a quantity that contains a mass and the square of a speed. Since the speed is essentially that of light, c, we might suspect that the desired ratio between mass and energy somehow contains the product Mc^2, with M representing the moving mass, that is, the mass dependent on the speed of the object.

I have been toying with the idea that energy might be given as simply the product, that is, $E = Mc^2$, or by some multiple or fraction of this expression, such as $E = \frac{1}{2} Mc^2$. As far as physical dimensions are concerned, this would be work: the expression Mc^2 contains the mass as well as the speed squared—except that we're now speaking of the speed of light. In any case, I can't get Mc^2 out of my mind. But since the M I use here is the mass of a rapidly moving particle, I can rewrite the equation with the help of the gamma factor:

$$E = m\gamma c^2.$$

The factor m is now the rest mass. I would surmise that this is the proper formula for very rapidly moving particles like the protons in the CERN accelerator, which have a large moving mass by comparison with their rest mass.

For small speeds the formula cannot be valid, since we know that my mechanics—I beg your pardon, Newton's mechanics—apply in that case; and according to those mechanics the kinetic energy of an object is simply $E = \frac{1}{2} mv^2$. This energy, as one would expect, disappears for zero speed. An object at rest has no energy according to my mechanics.

The above equation is a pretty odd one. Let's take it seriously for a minute, in the limiting case of the speed going to zero. I wouldn't get zero for the energy. The gamma factor for $v = 0$ is exactly 1, and that leaves me with the quite remarkable equation $E = mc^2$.

I must assume that this is a nonsensical equation. It would imply that there is energy even in an object at rest, and that this energy, measured by normal standards, is incredibly great, simply because the speed of light c is so great.

HALLER

Before we continue discussing this equation you call nonsensical, let me draw your attention to a small mathematical curiosity. Let's suppose we consider the formula $E = Mc^2$, which you mentioned previously, correct for all speeds.

NEWTON

Isn't that . . . ?

HALLER

Just a moment! Let me finish my argument. Let's calculate this relation for the case in which v is small compared with the speed of light c. A mathematical approximation is helpful here. Suppose x is a number much smaller than 1; then we can write:

$$\frac{1}{\sqrt{1-x}} \approx 1 + \frac{x}{2}.$$

This mathematical expression is not totally accurate; it holds only for sufficiently small values of x, and even for those it is not precise but only approximate. Take, for instance, $x = 0.02$. For the left-hand side of the equation, that yields the numerical value 1.0102. For the right-hand side, it yields 1.0100—so the relation actually holds quite well.

NEWTON

Of course, but what do you need it for?

HALLER

Very simply, I use this ratio to rewrite our energy equation. I replace x by $(v/c)^2$, which gives us

$$E = m\gamma c^2 = E = \frac{mc^2}{\sqrt{1 - \left(\frac{v}{c}\right)^2}} \approx mc^2\left(1 + \frac{v^2}{2c^2}\right) = mc^2 + \frac{1}{2}mv^2.$$

After taking a brief look at the equation I had written, my interlocutor jumped up in excitement, pointing at the paper.

NEWTON

But there it is, Einstein, my formula for energy: $\frac{1}{2}mv^2$. Wait, let me check that calculation once more. There's no doubt about it, it is correct. And now there is the mc^2 term. You know what that means, gentlemen? If the original equation $E = Mc^2$ is correct for all speeds, not merely approximate for very great speeds as I was suggesting a few minutes ago, then the energy of an object in slow motion consists of two parts—my old kinetic energy, given as the well-known $\frac{1}{2}mv^2$, plus a second term, the mysterious $E = mc^2$. If that's correct, it means my kinetic energy, $\frac{1}{2}mv^2$, represents just a minor correction; by far the greater part of an object's energy lies in its rest mass, given by mc^2, a huge quantity by comparison.

EINSTEIN

That isn't news to me, Newton. The total energy of an object is in fact given by the formula $E = Mc^2$. I derived this equation in my paper on the theory of relativity in 1905. The energy of an object with mass that is moving through space at a slow speed by comparison with c consists of the two parts you just mentioned. When the speed is zero, the energy of the object is certainly not zero, as your theory of mechanics would have it. Instead, it is given by $E = mc^2$, the product of its rest mass and the square of the speed of light.

NEWTON

But that's a huge amount of energy. Are you really saying that some objects at rest, like this pebble I'm holding, harbor a gigantic amount of energy?

EINSTEIN

That's precisely what I'm saying, Sir Isaac. According to my hypothesis, there is no fundamental difference between mass and energy. If I have understood Haller's earlier comment correctly, we still don't know why some particles have mass and others don't. As I see it, however, the mass of a particle is nothing more than "frozen energy."

Newton paced up and down, lost in thought. Then, full of excitement, he asked another question.

NEWTON

By using the term frozen energy, are you implying that this huge amount of energy—say, the energy frozen in the pebble I'm holding—might eventually be released simply by melting? You don't believe that, do you?

EINSTEIN

Conceivably that could happen.

NEWTON

Are you serious?

Newton suddenly looked pale and drawn, as though he hadn't slept for several nights.

EINSTEIN

If you don't believe me, ask Haller.

NEWTON (looking incredulous)

I hope it is clear to both of you: if this equation is correct, it leads to amazing consequences for our understanding of nature; it may also lead to catastrophic consequences for our planet. Who would guarantee that all the matter in our world is stable, that it might not dissolve into other forms of energy? It might suddenly change into light energy by means of a gigantic explosion.

Newton took a few steps up a hill that afforded a particularly attractive view of the lake and the Alps beyond. Einstein and I remained lying in the grass.

EINSTEIN

What does the scientific community think of my mass-energy equation today? Even though the equation assigns a very large amount of energy to every massive particle or body, it's not yet clear whether that energy can be released, either totally or partially. I'd love to find out more about it. But here comes Newton—I think we'll have to come to that point later.

NEWTON

Do you know what I've just been thinking? Consider the sun—we all know that it radiates vast amounts of energy day after day. Most of this is electromagnetic radiation energy. It has been going on for millions, maybe even billions of years.

HALLER

The sun has been in existence for more than four billion (4×10^9) years.

NEWTON

So much the better. At any rate, it has been radiating for a long, long time. Even while I was writing my book in Cambridge, I was wondering where all that energy might be coming from. If mass really does represent a kind of "frozen energy," the problem could easily be solved. We would simply suppose that certain processes occurring in the sun make use of that vast amount of "frozen mass energy." Of course, we would have to specify these processes. And should they exist, we would have to ask whether they could be made to happen on Earth, too. What a prospect! In that case, we would have an inexhaustible source of energy.

EINSTEIN

More than enough to blow the entire planet into the air—or, rather, to blow it into interstellar space.

NEWTON

Let's estimate how much energy your formula would predict for 1 kilogram (kg) of mass, using the common units of watt-seconds (Ws) and kilowatt hours (kWh). If this 1 kg mass moves at a speed of 1 meter per second (m/s), according to my theory of mechanics it has a moving energy of $\frac{1}{2}mv^2$, which makes ½ kg $(m/s)^2$, which makes ½ Ws. To calculate the energy term according to Einstein's equation, I just have to substitute c for v and multiply by 2:

$$1\,\text{kg} \times c^2 = 1\,\text{kg} \times (3 \times 10^8\,\text{m/s})^2 = 9 \times 10^{16}\,\text{Ws} = 25 \times 10^9\,\text{kWh}$$

$$[1\,\text{kWh} = 3{,}600\,\text{s} \times 1{,}000\,\text{W} = 3.6 \times 10^6\,\text{Ws}]$$

That's equal to about thirty billion kilowatt hours, an amount of energy that really defies the imagination.

HALLER

That's about the amount of energy produced by a very large power station with an output of 3 megawatts per year. A country the size of Switzerland could subsist on it.

However, I don't think there is any point in our philosophizing over the mass-energy formula. Even if Einstein's formula predicts that a comparatively large energy equivalent should be allotted to every unit of mass, that doesn't necessarily mean we can set all this mass free as energy. Since 1945, the end of World War II, a lot has happened that is connected to the conversion of mass into energy and so can be traced back to Einstein's formula. I think it would be best if we went through the possibilities in a systematic way.

EINSTEIN

Let's start right away, Haller. But you will understand that you will have to take the initiative in this discussion. Both Newton and I, well, you understand . . .

HALLER

I brought a copy of your 1905 paper with me.

EINSTEIN

You mean my second paper in the *Annalen der Physik,* the three-page article on relativity theory?

HALLER

Yes, that one. It's a short paper, but in view of its consequences not just for physics but for our whole understanding of nature and of the present state of our world, it may well be the most important scientific paper of the twentieth century. May I suggest that you read us the decisive conclusion in the final paragraph?

EINSTEIN (reading)

"If an object gives off energy L in the form of radiation, its mass is reduced by L/V^2 [where V is the speed of light (which we now always denote by c)]. In this process, it is of no consequence that the energy taken away from the object is converted directly into radiation energy; this leads us to a more general inference: The mass of an object is a measure for its energy content. If the amount of the energy changes by L, the mass is likewise changed by the amount $L/9 \times 10^{20}$ if we measure energy in units of ergs and mass in units of grams. It is not to be excluded that this theory can be tested, using materials with highly variable energy content—such as radium salts, for example."

HALLER

By the way, Sir Isaac, the erg is rarely used as an energy unit these days: $1 \text{ Ws} = 10^7$ ergs. Now back to our experiment. As we can see, even in his first paper on the energy-mass relation, Einstein predicted that it would be possible to test the theory by, for instance, a close study of the energy radiated by the element radium. And he mentioned an example where the importance of the energy formula immediately becomes obvious. I hope, Professor Einstein, you will allow me to present that example here, though in a somewhat modified form. When an object radiates energy in the form of electromagnetic waves, it steadily loses mass. A red-hot ball of steel will radiate electromagnetic energy in the form of thermal radiation. Suppose the sphere radiates an amount of energy E before it has cooled down to room temperature; according to Einstein's formula this amount of energy is equivalent to a mass E/c^2. Now, if I were to weigh this ball with great precision, both before and after the cooling off, I would have to find that it had lost weight in the process by the amount E/c^2. Unfortunately, the speed of light is so great that the difference in mass becomes minuscule and defies direct measurement.

Let me take another example: A 100-watt light bulb burning for one hour emits an amount of energy corresponding to a mass of 10^{-12} kilograms. Again, this is a tiny mass that cannot be directly measured.

Still, there are processes where the mass difference can be noticed right away, but all these are processes pertaining to atomic physics, nuclear physics, and particle physics. I would like to mention one of these in a little more detail. Why don't we look at the nucleus of heavy hydrogen?

NEWTON

I know what normal hydrogen is, but what's the heavy kind?

HALLER

There is a rare form of hydrogen with an atomic nucleus consisting not of a single proton as in normal hydrogen but of a bound system—a proton and a neutron. Neutrons, you may recall, are electrically neutral particles that behave like protons in many ways. One proton and one neutron can combine to form a new object that we call a deuteron. The electric charge of the deuteron is of course equal to that of the proton, which makes it possible to take a normal hydrogen atom and exchange its nucleus, the proton, with a deuteron. This new atom has a greater mass than normal hydrogen, since the deuteron is heavier than the proton. The scientific name for heavy hydrogen is deuterium. A very small fraction (about 0.016%) of the hydrogen that occurs naturally on Earth, that combines chemically in ocean water, for example, is in fact heavy hydrogen. Given the size of the oceans, even that small percentage makes a very large total amount of naturally occurring deuterium.

EINSTEIN

But why do a proton and a neutron combine to make up a deuteron? Could that be due to new forces?

HALLER

Certainly. There are very strong forces between proton and neutron that we call nuclear forces. A proton and a neutron attract each other very strongly when they are brought into proximity with each other: they then bind into a deuteron.

EINSTEIN

I think I can guess what you're driving at. Can you tell us something more precise about the respective masses of protons, neutrons, and deuterons?

HALLER (pulling a booklet from his pocket)

This booklet appears at regular intervals and is frequently updated. It contains a lot of information about elementary particles, including their masses. By the way, in particle physics, we don't measure the particles' mass in grams or milligrams; that would lead to very small numbers. Instead, we make use of Einstein's energy formula and determine mass in energy units, most often the electron volt units we discussed earlier.

NEWTON

What an interesting trick! Can you tell me what the mass of a proton is—specifically a proton at rest—in terms of electron volt units?

HALLER

Here's a table that shows the masses in electron volts, or rather in mega electron volts (MeV—1 MeV equals 1 million eV, or 10^6 eV) as well as in kilograms.

Proton mass	938.3 MeV = 1.673×10^{-27} kg
Neutron mass	939.6 MeV = 1.674×10^{-27} kg
Deuteron mass	1,875.7 MeV = 3.343×10^{-27} kg

Allow me to express the mass of my own body in MeV:

$$\text{Mass of Haller} = 4.49 \times 10^{31}\ \text{MeV} = 80\ \text{kg}.$$

EINSTEIN (having added up two numbers)

I thought so. Look here, Newton—since the deuteron consists of one proton and one neutron, we might expect its mass to be equal to the sum of the proton and neutron masses. But when you add these two masses, you get 1,877.9 MeV, which is 2.2 MeV, or 0.004×10^{-27} kg, more than the mass of a deuteron.

NEWTON

But how can that be? Didn't you tell us, Haller, that a proton and a neutron can be combined into a deuteron? Why isn't the mass of a deuteron equal to the sum of those two particles?

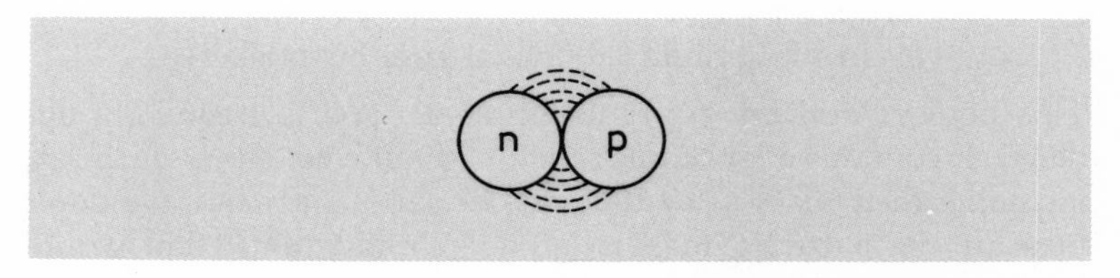

Fig. 15.1 The nucleus of a heavy hydrogen atom is called a deuteron; it consists of a proton, *p*, and a neutron, *n*, bound together by nuclear force. To separate them from each other, an energy of at least 2.2 MeV has to be applied, which is the energy corresponding to the mass difference between the deuteron and the sum of its two constituents.

HALLER

If you actually bring a proton and a neutron very slowly together under experimental conditions—say in a nuclear physics lab—you get more than a deuteron; in the process, energy is radiated off in the form of photons, electromagnetic radiation. This energy accounts exactly for the missing mass. Since the sum of the masses of its two constituents is slightly larger than the mass of the deuteron, we have what is known as a mass defect. It could also be called a mass deficit. So here we have a process that corresponds precisely to what Einstein described in his three-page paper: a system gives off energy in the form of electromagnetic radiation, and in the process it loses mass.

EINSTEIN

When a deuteron is formed, then, 2.2 MeV of energy are released, which is about 0.1% of the mass of the deuteron. This means that about one-thousandth of the original mass has been converted into energy.

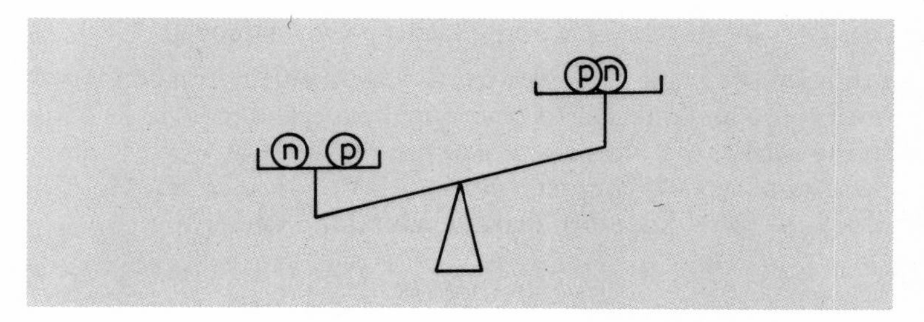

Fig. 15.2 The deuteron mass is about 0.12% less than the sum of the proton mass and the neutron mass; this difference is known as a mass defect.

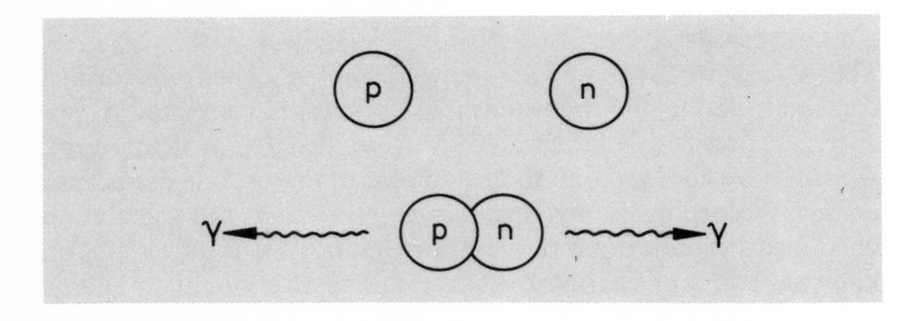

Fig. 15.3 When a proton is joined to a neutron, a system is created that emits energy in the form of photons. Typically, two photons of 1.1 MeV will be emitted in opposite directions. The sum of their energies corresponds to the mass defect described in fig. 15.2.

HALLER

And inversely, 2.2 MeV of energy must be added to the deuteron in order to break it into its constituents, a proton and a neutron. The procedure can be done quite easily in a laboratory—only it costs energy, as I said.

NEWTON

I really don't have a feeling for the amount of energy corresponding to 2.2 MeV, though I have no trouble visualizing energy in terms of 1 kilowatt hour.

HALLER

As a matter of fact, we can express the binding energy of the deuteron in terms of kilowatt hours as well. But I'm afraid that doesn't help us visualize it, since the amount turns out to be so tiny. My pocket calculator tells me the energy is almost exactly 10^{-19} kilowatt hours.

NEWTON

Let's suppose I have a large quantity of protons, neutrons, and electrons. I combine one proton and one neutron into one deuteron. I add one electron to the deuteron, so that I have constructed one atom of heavy hydrogen. Now if I proceed to make up a macroscopic amount of heavy hydrogen—a kilogram, say—how much energy is released?

EINSTEIN

That's easy to calculate. As we were saying, when we form a deuteron, about one-thousandth of its mass is converted to energy. Starting with a kilogram of heavy hydrogen (or deuterium), we will have changed one-thousandth of it, that is, one gram, into energy. According to my formula $E = mc^2$ and the calculation we did earlier, one gram of mass corresponds to about 25 million kilowatt hours of energy. And that enormous amount of energy would be released primarily in the form of electromagnetic radiation if we were to synthesize one kilogram of deuterium from neutrons and normal hydrogen.

Now suppose we were to carry out this procedure very rapidly. Wouldn't that give us a bomb of unimaginable destructive power? Doesn't that make you shudder, Haller?

HALLER

We shall have to talk about that possibility. At this point I would just like to point out that there is no naturally occurring matter that consists only of neutrons. That's why we can't produce macroscopic quantities of deuterium by using that procedure. So this is not the way either to produce energy or to make a bomb.

But look, we've been here quite a while. Do you want to stay longer, or shall we start back?

None of us felt like returning to Geneva, and so we decided to spend the evening in the Jura mountains. Since it was growing cool, Newton and Einstein walked into the forest to gather some dry wood, and I prepared a place for a fire. Before long, we had a comfortable campfire going. Newton and Einstein had brought enough wood to keep the fire crackling all evening.

SIXTEEN

The Power of the Sun

As we settled in front of the fire, basking in its warmth, the sun was close to the horizon in the west, its rays almost spent.

NEWTON (with a tinge of irony)

I suppose we have you to thank, my dear Einstein, for the fact that this fire gives us its warmth. The electromagnetic radiation it emits originates in the conversion into energy of a small part of the mass of the wood we put into it.

EINSTEIN (no less ironically)

I wouldn't look at it that way, Newton. The most important ingredient of our fire is the available combustible matter. Without the dry wood we managed to gather, there wouldn't be a fire. You can't feed a fire with rocks, although rocks have plenty of mass.

But my mass-energy relation speaks only of mass as such—that's to say, it concerns only one aspect of the situation, and a minor one at that. Whether we can ultimately convert mass into energy depends on other properties of matter, but you can hardly blame my equation for that.

NEWTON

Far be it from me! I'm still fascinated by the new horizons your mass-energy equation has opened up for me—and I'm merely scouring those horizons.

HALLER

Gentlemen, we should move on. There is nothing wrong with saying that some mass gets converted into radiative energy when the wood burns in our fire here; but that doesn't help much. The effect is really very small, and we can safely ignore it. It is true that the cinders that remain after the fire are lighter than the matter that's been burned, but the difference is only one part in 10 billion of the original mass. The accuracy with which chemists can today determine the masses we're dealing with is on the order of one part in 10 million. In other words, to be sensitive to the mass defect predicted by the theory of relativity, that accuracy would have to be improved by a factor of 1,000. Chemists are basically correct when they speak not only of energy conservation but also of mass conservation.

Whenever a substantial part of a particular mass is converted into energy, reactions in the area of nuclear or particle physics are involved—such as the production of deuterium in our previous example.

NEWTON

That would lead me to surmise that energy production in the sun is due to nuclear reactions.

HALLER

Research into the matter of which the sun and the stars are composed has shown that it consists largely of the noble gas helium. Approximately a quarter of the matter of the universe is simply helium. A helium atom consists of two electrons in the outer shell, and a nucleus which is itself made up of two protons and two neutrons. This nucleus is called an alpha particle, designated by the Greek letter α.

If we look at this nucleus more closely, things become interesting; expressing its mass in units of energy, as Einstein's equation suggests, we get

$$m_\alpha = 3{,}727.5 \text{ MeV}.$$

NEWTON

What a strange nucleus! I have just compared its mass with the sum of the masses of its constituents, in other words, with two proton masses plus two neutron masses. Those masses total 3,755.8 MeV, which is 28.3 MeV more than the mass of the alpha particle. The difference is a fairly significant energy or mass: It makes up almost 0.8% of the entire mass of the alpha particle. And this energy is released when two neutrons combine with two protons to form an alpha particle. Yet the deuteron's binding energy was only 2.2 MeV, which is only 0.1% of the mass of deuteron.

EINSTEIN

That must mean that the helium nucleus is a relatively stable structure, with its four constituents bound strongly together.

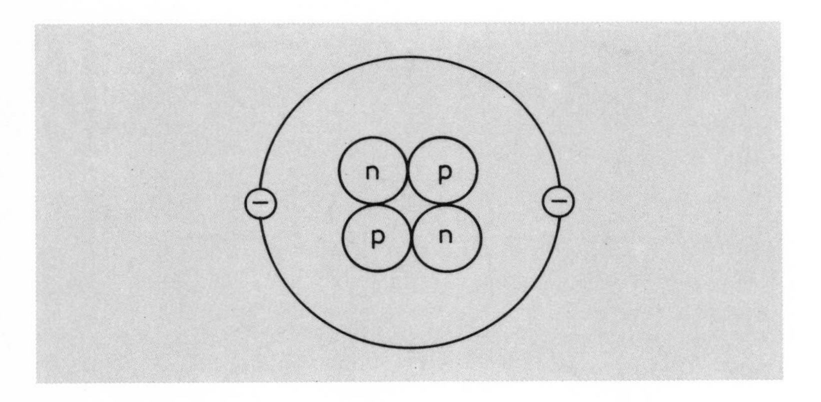

Fig. 16.1 Schematic representation of a helium atom: a shell of two electrons surrounds its nucleus, which consists of two protons and two neutrons. The size of the nucleus is greatly magnified in the diagram.

HALLER

This characteristic of the helium nucleus is due to specific properties of the nuclear forces, which we can't go into right now.

The alpha particle can be considered a bound system of two deuterons. The mass defect due to the binding of these two deuterons is 23.9 MeV, somewhat less than the 28.3 MeV mentioned by Newton.

NEWTON

So if I combine two deuterons, I release 23.9 MeV of energy. That would certainly be an interesting way to form an alpha particle while acquiring a significant quantity of energy. You told us earlier that there is no lack of deuterons on Earth since the oceans contain plenty of heavy hydrogen. So we could follow a recipe: take two deuterons, put them in a pot, stir—and lo and behold, out comes a helium nucleus plus a lot of radiation energy!

HALLER

On the face of it, you're correct. The combination of two deuterons, also called the fusion of two deuterons, releases ten times as much energy as the fusion of a proton and a neutron. If we were to produce one kilogram of helium by fusing deuterons, we would end up with 200 million kilowatt hours of energy—a considerable amount.

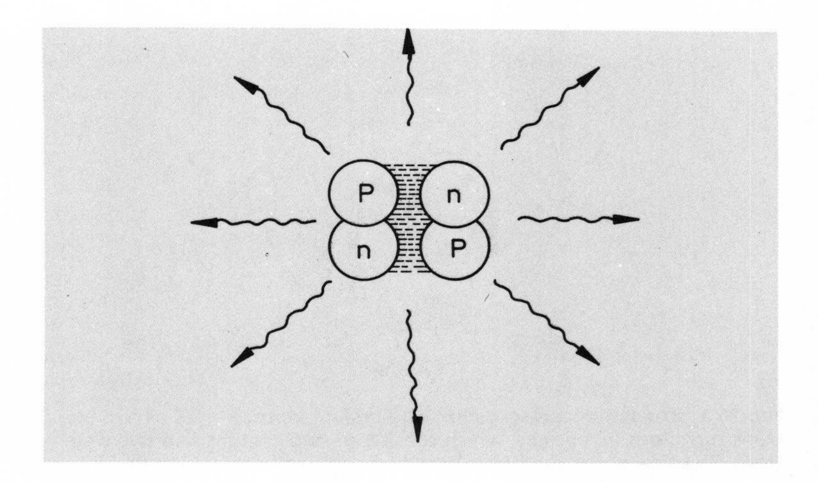

Fig. 16.2 When the deuterons fuse to create a helium nucleus (also called an alpha particle), about 24 MeV of energy is released. It will be radiated off in the form of photons.

That may sound simple, but it certainly isn't. The problem is that the deuteron, like the proton, has a positive electrical charge, which means that two deuterons will repel each other electrically. In order to melt them into an alpha particle, we must overcome this repulsion and bring the deuterons into such close proximity that the nuclear forces ultimately take over.

EINSTEIN

At what distance does that happen?

HALLER

At about 10^{-12} cm—that is, approximately one ten-thousandth of the diameter of an atom.

EINSTEIN

Well, good luck! The probability that two deuterons will approach each other that closely when we shoot them toward each other must be minimal—and the more so if we have to overcome the electric repulsion between the two particles.

HALLER

Therein lies the problem. It's easy enough to achieve fusion once in a while by shooting deuterons at each other. But the energy of the particles must be sufficiently high to overcome the repulsion. Only very few deuterons actually take part in the fusion process; most simply bypass each other. These processes have been carefully studied; and it turns out that the energy needed to accelerate the particles sufficiently to initiate fusion is much greater than the energy that can be produced by the occasional occurrence of fusion. In other words, there's no net gain in energy; energy is actually lost.

EINSTEIN

That's what I thought from the start. But I've just had another idea. Suppose we heat deuterium up. Heating material up merely means an increase in the kinetic energy of its constituent atoms or molecules. And at a certain temperature, probably a very high one, a few deuterons will manage to overcome the electric repulsion during their frequent collisions, and will fuse together; at still higher temperatures, many will do so. In other words, if deuterium is hot enough, fusion will automatically be initiated; a sort of nuclear burning will begin. This process might even lead to an explosion if enough deuterium were available.

NEWTON

Could that be the process responsible for producing energy in the sun?

HALLER

One thing at a time, please! Einstein's idea is certainly correct. Nuclear fusion will set in automatically at a certain temperature, but unfortunately a very high one in relation to normal temperatures found on the Earth's surface. To overcome the electrical repulsion between the deuterons, the particles involved must have an energy of several MeV. But to achieve an average deuteron energy in the range of several MeV would require temperatures on the order of 10^{10} degrees, that is, 10 billion degrees. And that's a temperature beyond our imagination.

Of course, to get the fusion process started, not all the deuterons have to have the required energy. Even at significantly lower temperatures, a small percentage of the deuterons will have enough energy. A temperature of about 10^8, 100 million degrees, would be enough to initiate fusion.

NEWTON

It isn't quite clear to me what heating up substances such as heavy hydrogen to a temperature of a million degrees would imply. What happens to the atoms at such enormous temperatures?

HALLER

The answer is quite simple—atoms as such cease to exist. As soon as the energy of the colliding atoms exceeds the order of 10 eV while the deuterium is being heated up, the electrons will be kicked out of their shells. The energy needed to remove an electron from an atom is of about this magnitude. For normal hydrogen, for instance, it's 13.6 eV.

So you can see that at temperatures above a million degrees there will be no more atoms of heavy hydrogen. Instead, we have a highly heated mixture of deuterons and electrons, known as plasma. If we produce plasma in this way and heat it to temperatures slightly below 100 million degrees, fusion will set in. Thermonuclear combustion will have started.

NEWTON

All right. Back to the sun, then. From all you've said so far, I surmise that the sun produces its energy by the thermonuclear burning of heavy hydrogen.

HALLER

That's close to the truth. The temperatures inside the sun are high enough for fusion to occur, and so a significant fraction of mass gets converted into radiation energy according to Einstein's $E = mc^2$. The energy is acquired through helium synthesis. But the process we've just been examining—the synthesis of helium nuclei via the fusion of two deuterons—makes up only a small fraction of the energy produced by the sun. The major part of the sun's energy derives from a more complicated process that occurs in several stages and involves the synthesis of helium nuclei out of protons, the nuclei of normal hydrogen.

EINSTEIN

A helium nucleus contains two neutrons. If helium is produced from normal hydrogen, with a nucleus consisting of only one proton, where do those neutrons come from?

HALLER

As I said, the reaction proceeds in several stages, in the course of which neutrons are formed from protons. We now know that protons and neutrons are closely related, and a proton can be converted into a neutron. Other particles, especially electrons and neutrinos, play a role in this process. But we needn't go into detail here. The point is, even normal hydrogen can be converted into helium in due course.

NEWTON

I assume it is not especially difficult to estimate how much energy the sun radiates in a time period of, say, one second. From this amount we could calculate how much mass the sun loses per second.

HALLER

That's easily done. If I remember correctly, the sun's energy output is approximately 3.7×10^{23} kilowatts, of which only a minute fraction reaches our planet in the form of electromagnetic radiation. That amounts to about 10^{11} kilowatts—about 100,000 times more energy than that produced by all the power plants on Earth. From the energy loss we can use Einstein's equation to calculate the amount of mass lost by the sun per second: four million tons.

NEWTON

That's a lot. Considering that the sun has been sustaining this loss over several billion years, one wonders how long it can go on.

HALLER

Don't worry. The sun can survive the loss for another few billion years without any problems. One thing is clear, however: the sun and the stars shine only because they are capable of converting mass into radiation; and the energy balance is dictated by Einstein's equation. Without radiation of the sun's mass via the thermonuclear combustion of hydrogen, there would be no energy from the sun and thus no life on Earth.

NEWTON

So life would not be possible in the universe without the effects of relativity. There would be nothing but cold matter.

EINSTEIN

You exaggerate, my dear Newton. As Haller pointed out earlier, my equation determines only the energy balance. The possibility of nuclear fusion is not a result of my equation but a consequence of the specific nature of nuclear forces. It's similar to the way things work in a bank: my equation ensures that the balance is correct and that the accounting department doesn't make any mistakes. Where the money comes from is another, more difficult problem.

NEWTON

But let's get back to Earth. Whatever the sun is capable of should in principle be possible for man to imitate on Earth. Is it possible to achieve nuclear fusion on Earth?

HALLER

You'll understand later on why your question can't be answered with a simple yes or no. I suggest we first discuss another kind of energy production by means of a nuclear process—the *fission* of atomic nuclei. Here again, the energy balance is given by Einstein's equation.

EINSTEIN

Agreed, though I don't understand how we can gain energy by splitting atoms, no matter how we go about it. Didn't we just figure out that we can produce a lot of energy by joining two deuterons together so that they form a helium nucleus—the process called fusion? Of course, I can invert the process by splitting the helium nucleus into two deuterons, but that will cost me a fair amount of energy and will result in no net gain.

HALLER

That's true if it's helium you're trying to split. The constituents of the helium nucleus have quite a high binding energy. But binding energy is specific to the structure of the nucleus in question. It's dependent on the number of protons and neutrons in the nucleus.

NEWTON

Are you telling me that the binding energy of individual nucleons is even greater for atomic nuclei that are heavier than helium?

HALLER

Yes, it can be greater. The greatest binding energy is found in the nuclei of iron. The nucleus of an iron atom is the most stable atomic nucleus, and that's why we have so much iron on Earth.

EINSTEIN

I take it, then, that atomic nuclei heavier than iron nuclei, such as the nuclei of lead or gold, are less stable than those of iron.

HALLER

That's right, and this phenomenon is easy to explain. We know that very heavy nuclei contain many protons. The atomic nucleus of gold, for instance, has 79 protons, and all have a positive electric charge: they repel each other. If we were to switch off the nuclear forces all of a sudden, the nucleus would explode. The protons would all fly off at tremendous speeds.

We can't increase the number of protons in a nucleus at will. The electrical repulsion between protons becomes more and more important as the nuclear size increases, and ultimately it leads to instability in the nucleus. At the slightest disturbance from outside, the nucleus will break up. It might be split in two, for example.

EINSTEIN

I see. So we may expect an atomic nucleus above a certain size to have the tendency to split in half; in this process energy will be released simply because each half contains fewer protons than the initial nucleus. And that means they are more strongly bound.

NEWTON

Are there any nuclei that can spontaneously split in half?

HALLER

The best-known one is the uranium nucleus. It has 92 protons and usually, but not always, 146 neutrons. Very occasionally a uranium nucleus splits spontaneously, yielding lighter nuclei with fewer but more strongly bound protons. As a specific instance, a uranium nucleus may split up into a barium nucleus with 56 protons and a krypton nucleus with 36 protons.

Although fission in a uranium nucleus rarely occurs spontaneously, it can easily be induced with a little help from outside. If

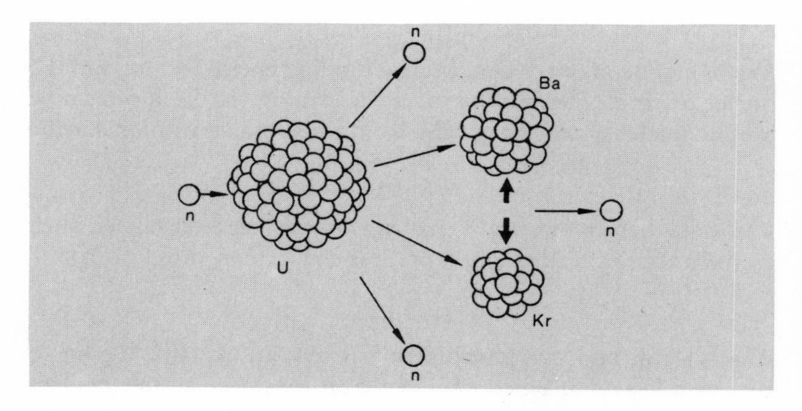

Fig. 16.3 At left, a neutron collides with a uranium nucleus, which contains 92 protons and (usually) 146 neutrons. The incoming neutron transfers its energy to the nucleus and thereby excites the resulting system such that it pulsates like a soap bubble, and finally breaks up into two smaller nuclei and several neutrons. Here it is shown breaking up into a barium nucleus with 56 protons, a krypton nucleus with 36 protons, and several neutrons.

you shoot a neutron at a uranium nucleus, imparting energy to the nucleus and heating it up, it will start oscillating like a big soap bubble and will finally burst. The energy of the impinging neutron doesn't even have to be very great—the nucleus needs only a small jolt to split. The analogy with the soap bubble is pretty good. The bigger the bubble, the greater the tendency to split into smaller bubbles, if not to blow up altogether.

EINSTEIN

How much energy is released if a uranium nucleus is split?

HALLER

About 200 MeV. But this energy must be considered in relation to the energy frozen in the heavy uranium nucleus according to your equation; compared with that, only 0.1% of the uranium mass has been converted into energy.

NEWTON

As far as the transformation of mass into energy is concerned, this fission process seems to be much less effective than the fusing of deuterons into helium, where almost 1% of the mass is changed into energy.

HALLER

Nevertheless, even with fission a relatively large fraction of the mass is transformed into energy—at least by comparison with our campfire here. But nuclear fission yields less electromagnetic radiation than the fusion of helium nuclei does. Most of the energy generated in fission goes into the kinetic energy of the two resulting nuclei: they go racing off at huge speeds.

And it mustn't be forgotten that fission leaves us not only with two smaller nuclei; a number of neutrons are also emitted.

NEWTON

But why is it so important to have free neutrons among the leftovers?

HALLER

It's not important for fission as such; that could easily occur without any leftover neutrons. But if we want to initiate nuclear fission on a large scale, we can't just work with an individual nucleus but have to split a great number of nuclei in a short time. And that can't be arranged from outside. There has to be some help from the fission process itself.

NEWTON

Ah, I understand. The surplus neutrons fly around in the uranium matter and induce further fission processes.

HALLER

Right. That may lead to a chain reaction not unlike the chain reaction we observed when I lit this fire. First I lit a small piece of paper, then the flames took over some dry foliage and a few small twigs, and finally the whole set of logs was ablaze.

By the way, it isn't only in uranium that a nuclear chain reaction can be initiated. It can happen in other elements—plutonium, for instance, which has 94 protons in its nucleus.

EINSTEIN

When a chain reaction sets in, it appears to lead to a sort of burning of nuclear matter. I could imagine these reactions accelerating and building up into a real nuclear explosion.

HALLER

To start with, you need enough fissionable material to allow most of the neutrons that are produced to start new fission processes.

We use the term "critical mass." The critical mass for the uranium isotope 235 amounts to about 50 kilograms. This isotope is a particular type of uranium with a total of 235 nucleons in its nucleus—143 neutrons in addition to the 92 protons.

EINSTEIN

How big a chunk of uranium would that be?

HALLER

It would be a ball of uranium with a 17-centimeter diameter—about the size of a soccer ball.

EINSTEIN

Does that mean that if I had a ball of uranium here, it would immediately blow up, converting 0.1% of its matter, or 50 grams, into energy?

HALLER

Yes it does. There would be a huge explosion. That's how the atomic bomb was built. And you've heard about the incredible destructive potential of that weapon. It's just as well we don't have a chunk of uranium with us here.

Newton had jumped up and was pacing up and down in front of our campfire. He now stopped and spoke.

NEWTON

I suspected that. As soon as you hinted that it's possible to release a small fraction of the energy Einstein's formula tells us is frozen into all matter, I knew. So that's what nuclear weapons are about.

HALLER

Yes, those are the kind of weapons used over Japan in 1945, at the end of World War II. They are often inaccurately called atomic weapons. Around 1940, when the things we've been discussing this evening became known in the scientific community, a group of physicists in the United States—including you, Einstein—feared that scientists in Nazi-dominated Germany might be able to develop a nuclear bomb. That was well before the end of the war, which was wreaking havoc in Europe at that time. The American physicists turned to President Franklin Roosevelt, and he, along with the high command of the armed forces, started a special program for the construction of nuclear weapons. It's quite a long story, but I'd like to go into it in some detail.

EINSTEIN (with furrowed brow)

I hope you will. And please don't leave anything out.

HALLER

Last year I spent several weeks in Los Alamos, New Mexico. That's where the first bomb was built. While I was there, I got to know a physicist who was actively involved in what was code-named the Manhattan Project. One day I took a hike with him through the canyon country outside Los Alamos, and he told me many details about what happened at that time. I'll try to give you a shortened version of what he said.

SEVENTEEN

Lightning at Alamogordo

Newton and Einstein were fascinated with my report of the Manhattan Project. I began with the origin of the project, which first took shape, not long after the outbreak of World War II, in the minds of a few leading physicists and some influential military people in Washington. I briefly described the career of J. Robert Oppenheimer, the brilliant young physicist who later became director of the project when it reached its culmination in Los Alamos, on the high mesas of New Mexico. The goal of the Manhattan Project was the production of several atomic bombs in the shortest possible time. The principal difficulty lay in procuring fissionable material, which had to be extracted from uranium in a complicated procedure. It was only after the war in Europe had ended that it finally became possible to carry out the first test explosion in the New Mexico desert.

Jornada del Muerto, "A Day's March of the Dead," was the name given by the Spanish conquistadors, four hundred years ago, to the barren stretch of land that today lies in the state of New Mexico, south of the city of Socorro. The name was appropriate, for it is a particularly dry and forbidding part of the desert. It was here, not far from Alamogordo, that Oppenheimer had chosen the spot where the first trial explosion was to take place. Its location was to enter historical accounts under the name "Trinity."

Engineers erected a steel support tower 30 meters high, so that the nuclear payload could be detonated above ground level. In this way, the crater below the explosion would be minimal, and there would be no great mushroom of dust rising into the skies.

The test was carried out on July 16, 1945. The assembling of the equipment had started days before. The important part of the payload was a complicated mechanism that would allow a very rapid implosion of the fissionable material, caused by conventional explosions.

On July 16, just after 5 A.M., Oppenheimer and General Groves, the military commander of the project, took up their position in the observation bunker. Shortly thereafter the explosive device was ignited, and that was the start of the nuclear age. The first seconds after the explosion were described by the physicist Otto Robert Frisch as follows:

> And then without a sound, the sun was shining; or so it looked. The sand hills at the edge of the desert were shimmering in a very bright light, almost colourless and shapeless. This light did not seem to change for a couple of seconds and then began to dim. I turned round, but that object on the horizon which looked like a small sun was still too bright to look at. I kept blinking and trying to take looks, and after another few seconds or so it had grown and dimmed into something more like a huge oil fire, with a structure that made it look a bit like a strawberry. It was slowly rising into the sky from the ground, with which it remained connected by a lengthening stem of swirling dust; incongruously, I thought of a red-hot elephant standing balanced on its trunk. Then, as the cloud of hot gas cooled and became less red, one could see a blue glow surrounding it, a glow of ionized air. . . . It was an awesome spectacle; anybody who has ever seen an atomic explosion will never forget it. And all in complete silence; the bang came minutes later, quite loud though I had plugged my ears, and followed by a long rumble like heavy traffic very far away. I can still hear it.

Robert Oppenheimer later remembered this extraordinary moment: "A few people laughed, a few people cried, most people were silent. There floated through my mind a line from the Bhagavad-Gita in which Krishna is trying to persuade the Prince that he should do his duty: 'I am become death, the shatterer of worlds.' "

In addition to the test device, two more atomic bombs had been produced in Los Alamos. The first was dropped by order of President Harry Truman on August 6, 1945, on the Japanese harbor city of Hiroshima. Three days later, the second bomb was detonated above Nagasaki. The two bombs killed more than 100,000 people.

Thus ended my narrative. Newton had closed his eyes. Einstein stared into the flames. Before long, I heard him repeat: "Now I am death, the destroyer of all worlds."

NEWTON (putting his hand on Einstein's shoulder)

Chin up, Einstein. Those physicists in Los Alamos didn't invent nuclear fire; all they did was drag it down to Earth from the sun.

It was just a matter of time before it had to happen. I would bet that any civilization that develops in any corner of the universe will at some point be capable of doing likewise and will act accordingly. Whether this energy source is used to build bombs is, of course, another question. It will ultimately be answered not by physicists but by politicians and the people who elect them.

HALLER

I should mention that the majority of the physicists involved in building the first atomic bomb voted in favor of exploding the bomb above an uninhabited stretch of land. It was to serve as a demonstration, a warning to the enemy. It was the politicians who decided otherwise, against the advice of the scientists. Let me quote from a letter that you, Mr. Einstein, wrote in 1950: "I've never taken part in any enterprise of a military or technical nature, and I've never done any research that was directed towards the production of atom bombs. My only contribution to this whole business is that I determined the mass-energy relation in 1905. This is a fact of physical nature. It is very general, and its potential connection to military applications couldn't be further from my thoughts."

EINSTEIN

I always was a pacifist, no doubt about it. I would never work on producing bombs. I agree with you, Sir Isaac, that it is just a matter of time before a developing civilization will discover the possibilities that open up with nuclear fission and nuclear fusion. But I also ask you: Will that civilization, or will our human civilization here on Earth, be able to survive with this knowledge? Or will it finally succumb to the perpetual temptation to play with nuclear fire?

HALLER (answering instead of Newton)

Nobody knows whether we will be able to resist that temptation in the long run. Maybe it's the fate of every civilization in the universe finally to burn in the same nuclear fire that makes life possible in the first place. But then again, maybe we'll succeed in avoiding the nuclear inferno and thereby ensure the survival of our planet. Nobody knows the answer to your question, Einstein. But one thing is certain: at that zero hour in 1945 when the first nuclear explosion occurred, a new age began—an age in which armed conflict between nations that have nuclear weapons at their disposal can no longer be considered a practical means for the resolution of differences. So I see the construction of nuclear weapons as a challenge to solve conflicts in ways other than war. I'm quite optimistic that this will happen.

EINSTEIN (looking up)

Do you really believe that, Haller? Do you really assume that the leaders of a nation would renounce the use of nuclear weapons when push comes to shove? You must be quite an optimist. Maybe you're right. I certainly hope so.

One thing has become clear to me this evening: The equivalence of mass and energy that I discovered while working at the Bern patent office has played its part in releasing the forces inherent in atomic nuclei—and in this sense, my equation probably has changed the world. What the equation hasn't changed is our way of thinking and our way of solving conflicts. I think we need a new way of thinking so that humanity won't perish by nuclear forces. But where should this new way of thinking start if not with the people who brought thermonuclear fire to Earth—physicists, scientists, and technicians? I do think, Haller, that history has put a heavy responsibility into your hands and those of your colleagues.

HALLER

After the conclusion of the Manhattan Project, Oppenheimer left Los Alamos. At the ceremony arranged at his departure he made a short speech, ending with the following words:

> If atomic bombs are to be added as new weapons to the arsenals of a warring world, or to the arsenals of nations preparing for war, then the time will come when mankind will curse the names of Los Alamos and Hiroshima. The peoples of this world must unite, or they will perish. This war, that has ravaged so much of the earth, has written these words. The atomic bomb has spelled them out for all men to understand. Other men have spoken them, in other times, of other wars, of other weapons. They have not prevailed. There are some, misled by a false sense of human history, who have told us that they will not prevail today. It is not for us to believe that. By our works we are committed, committed to a world united, before the common peril, in law, and in humanity.

Today we must do everything we can to make sure that our planet won't be destroyed by radiation due to our own imperfection and stupidity. I side with Oppenheimer as far as our knowledge is concerned, including knowledge about the danger. That knowledge might defeat the danger.

Still, the question remains: Will our knowledge be strong enough? For most people, Einstein's equation doesn't describe a profound element of nature, an element responsible for our very existence. It's regarded as a magic formula invented by physicists, like the atomic bomb and the hydrogen bomb—something we can't get rid of, something to be afraid of. But fear is not a good signal to follow.

We must also clarify what we mean by knowledge in this context. We must distinguish between intelligence and reason. Our intelligence allows us to view the world in terms of its own laws, irrespective of us. With the help of intelligence we have built up a rational edifice of scientific knowledge, forever developing with no apparent limits, no longer open to a complete overview.

Reason, on the other hand, allows us to set limits if necessary, limits that shouldn't be crossed. Reason is the authority that includes human beings with all their imperfections and all their limitations. It isn't restricted to the cold, rational world of scientific knowledge, which carries no measure in itself. Will reason ultimately prevail? I don't know—nobody knows.

Einstein got up and looked up at the stars. There was the Milky Way, stretching its band across the firmament. He turned to Newton.

EINSTEIN

Sir Isaac—what an evening! You, in Cambridge, and I, in my youth, wondered why those stars shone. How often, as a boy, did I climb to the attic of our house in Munich to look at the stars and ask myself that very question. Now we know why the stars and the galaxies light up the dark universe, and why the sun's light warms the Earth. When I wrote those three pages for the *Annalen der Physik* in Bern in 1905, the pages that contained my equation $E = mc^2$, I never dreamed that I might have found a key to the inexhaustible energy supply hidden in the stars. An equation that ultimately, together with the equations of nuclear physics, would create the theoretical basis for new weapons more deadly than anything ever invented in their potential for destruction.

That's the way it goes in physics: we build up theories that we don't really take seriously at first. When I wrote my equation, no one could have conceived these consequences—I couldn't, nor could Max Planck in Berlin or Arnold Sommerfeld in Munich. Reality caught up with us and overtook us. I think we should not only take our theories seriously but should take them seriously beyond what the moment appears to dictate. And we should always remember that our knowledge is not just for us—the public has to be informed fully and truthfully. Science is a serious business—too serious to be left solely to scientists.

EIGHTEEN

Energy Hidden in the Nucleus

Next morning I met Einstein for breakfast on the terrace of the CERN cafeteria. He looked tired after our late return from the mountains and greeted me a bit gruffly. He drank his cappuccino in silence. Newton, on the other hand, was in high spirits and ready for action when he finally appeared. He had already taken the time to walk around the CERN grounds.

NEWTON

On my morning walk I saw a lot of things I would like to ask you about. But we agreed yesterday that our discussion should continue in a systematic manner. So let's first talk about the ways in which nuclear energy is being used today.

HALLER

Since the explosion of the two atomic bombs at the end of World War II, no other nuclear bomb has been dropped over a populated area; on the other hand, there's been intensive work on the development of new weapons. The most significant advance in this respect has been the building of a bomb based on nuclear fusion.

EINSTEIN

I barely got any sleep last night; I couldn't help imagining all that could be done with uranium or plutonium bombs. One thing that occurred to me was that the explosion of a uranium bomb could be used to generate, for a very brief time, the high temperatures that are required to initiate nuclear fusion. I'm afraid that would be possible for use only as a bomb—and I mean a hydrogen or deuterium bomb. But this would be a great deal more destructive than the uranium bomb.

HALLER

That certainly comes to mind. In fact, right after World War II, several experts in the United States and the Soviet Union were working toward a hydrogen bomb along these lines. And they succeeded remarkably quickly, only ten years after the start of the Manhattan Project. The way it works is simple. A small atomic bomb explodes and raises the fuel to the temperature needed for fusion. The explosion of a fission bomb initiates a second, much more powerful explosion.

In the past decades, the superpowers—especially the United States and Russia—have been amassing a horrendous arsenal of nuclear devices, of various kinds but all built according to this principle. The effectiveness and hence the destructive potential of a hydrogen bomb are far greater than those of an atomic bomb based on fission. There is no doubt that if a hydrogen bomb were ignited on Earth, the resulting light flash would be visible to the naked eye from as far away as the moon.

A large hydrogen bomb has the destructive power of about 200 tons of TNT, which, as you probably know, is a very effective explosive. About half the power generated in the explosion goes into the tremendous pressure wave that starts from the point of the explosion; about one-third appears as heat and light radiation issuing from the explosion.

You can imagine what would happen if a bomb of that kind were to be ignited in a clear sky about 50 kilometers above the city of Geneva. The city would be destroyed, and so would the small towns around Lake Geneva and the surrounding French countryside. In an area of more than 100 kilometers around Geneva, as far as the city of Bern, and way into the Juras and the Alps, all the forests, all flammable material, would burn. In other words, all of western Switzerland and the neighboring French provinces would be laid waste. There would be about one million human victims. If a hydrogen bomb were to explode above a densely populated area, such as the Ruhr district in Germany or the cities of New York and Moscow, there would be millions of human victims.

NEWTON

Let me plead that we abandon this atrocious chapter on the "utilization" of nuclear energy. I'm much more interested in the possibility of producing energy from atomic nuclei for peaceful purposes—whether by fusion or by fission.

HALLER

Let's first talk about fusion, the method of energy production that occurs inside the sun. And when I speak of nuclear fusion, I'm not only speaking of the fusion of deuterons into helium nuclei. Another interesting process is the fusion of deuterons and tritons.

A triton is a nucleus that has one proton and two neutrons. It can be created by adding a neutron to a deuteron. If we now add an electron, in such a way that it orbits around this nucleus, we obtain an atom of superheavy hydrogen, or tritium.

By fusing a neutron with a triton, we obtain a helium nucleus and a neutron:

$$d + t \rightarrow \text{He} + n\,.$$

This reaction can be formulated differently by indicating the individual initial nucleons:

$$(p + n) + (p + 2n) \rightarrow \text{He} + n\,.$$

We start with two protons and three neutrons and end with the same number. The nucleons simply change their partners. But since helium is a very tightly bound nucleus, this reaction actually generates energy—kinetic energy. In other words, the resulting helium nucleus and the neutron are moving away at great speed from the site of the interaction.

Analyzing this process in detail, we find that 0.4% of the initial mass is transformed into energy. This reaction, by the way, is the one mainly used for thermonuclear weapons. Obviously, we should ask ourselves whether we could also use it—or, for that matter, any other fusion reaction—to produce energy for peaceful purposes. But that hasn't yet proved possible despite our best efforts.

EINSTEIN

I suppose it's because of the difficulties involved in heating the fuel, deuterium or tritium or whatever, to a temperature of about 100 million degrees.

HALLER

Temperatures of more than 10 million degrees but significantly less than 100 have been reached for very brief periods, fractions of a second. But that's not enough. Whether we'll ever get there is written in the stars, if you'll permit the pun: you might say it's written in the place where fusion actually occurs.

NEWTON

But how have those temperatures been reached here on Earth?

HALLER

First, the fuel is heated to approximately 12,000 degrees. That's enough to strip the electrons from the nuclei and to change the atomic fuel into plasma. Next, the plasma gets strongly compressed by intense magnetic fields, which heats it even further; about 40 million degrees have been reached by this means. But, as I said, that has been possible for only very brief periods.

At present, the most advanced lab in Europe for this kind of research is the JET (Joint European Torus) Lab in England—at Culham, near Oxford. Scientists there are inching their way toward the temperatures needed for nuclear fusion and have recently made some progress in that direction.

Other recent attempts to reach the required temperatures have been made by shooting photons—or, more accurately, laser beams—into the fusion fuel. A powerful laser beam can heat up deuterium and create the needed fusion temperature for a brief time. Millions of fusion reactions have been initiated like this; but a chain reaction of thermonuclear combustion—which we hope would remain a controlled one—has not yet resulted. Whether we'll ever manage to build power stations based on fusion for usable energy, we don't know. But we do know this: if it ever does happen, there will be no limit to the energy we could generate. There's a plentiful supply of fusion fuel, especially deuterium, on Earth. All the energy needed by a country as large as the United States for one day could easily be generated by the fusion of a mere 250 kilograms of deuterium and tritium.

EINSTEIN

But what about nuclear *fission*? How far has that been employed for peaceful energy production?

HALLER

The technical aspects of controlled nuclear fission look a great deal more favorable at this date than those of nuclear fusion. And that's easy to explain: nuclear fission occurs spontaneously; there's no need for the fissionable material to be brought to a very high temperature or otherwise treated in some particular way.

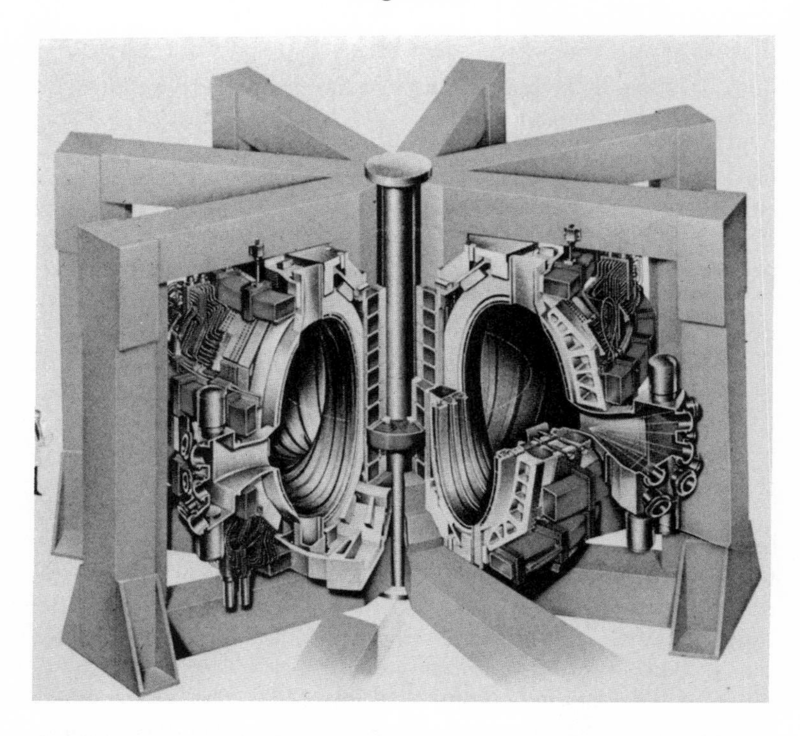

Fig. 18.1 The European fusion experiment JET (Joint European Torus) near Culham, England. (Schematic image courtesy of JET Joint Undertaking, Culham.)

For another thing, it's easy to control the fission process. We have already seen that for a given amount of a fissionable material, such as uranium, a chain reaction will set in if there is enough of that substance. A critical mass needs to be present. In a nuclear reactor, fission is controlled by an appropriate steering mechanism so that fission doesn't proceed like an avalanche or a bomb explosion, but occurs in a continuous, stable fashion. For this purpose we need precise control over the number of neutrons bouncing around in the reactor and constantly initiating new fissions. This control can be exercised by means of rods consisting of material whose nuclei can easily absorb those neutrons—cadmium, for example. As soon as the neutrons are introduced into the uranium core, the chain reaction will stop. Retract them a bit, and fission will slowly recur. So fission can be controlled, and the reactor can be shut down as soon as anything goes wrong.

Fig. 18.2 Photograph of JET experimental setup. (Courtesy of JET Joint Undertaking, Culham.)

Fig. 18.3 Interior of the annular combustion chamber of JET. In this ring, plasma is heated to a temperature of several million degrees by means of strong magnetic fields. The scale is illustrated by the technician at left. (Courtesy of JET Joint Undertaking, Culham.)

Fig. 18.4 A view of the inside of ASDEX combustion chamber. In this fusion experiment at Garching, near Munich, Germany, plasma was heated to more than 10 million degrees. Shown here is a hydrogen pellet, which has just been injected from the right while still frozen, in the process of evaporating. (Courtesy of Max Planck Institute for Plasma Physics IPP, Garching, Germany.)

EINSTEIN

Still, I can't help thinking that this procedure is a bit risky. There might be some unexpected combination of unlikely events, and the whole thing could blow up, don't you think?

HALLER

You certainly don't have to worry about a nuclear explosion like an atomic bomb. A reactor is built very differently from a bomb. Even if things go seriously wrong and all the controls fail, the worst thing that can happen is a meltdown in the reactor. There will never be a nuclear explosion; a nuclear reactor simply does not have the critical amount of fissionable material of the concentration needed for a bomb.

In this connection I should point out that the first reactor on our planet was constructed not by humans but by nature. Years ago, a remarkable concentration of a particular uranium isotope was found in the uranium deposits of Oklo, in the West African country of Gabon. The only explanation for this strange phenomenon was that the uranium concentration was the remnant of a natural nuclear reactor. Experts calculated that 1.8 billion years ago a nuclear chain reaction must have occurred in this location that was sustained for as long as a billion years. It was even possible to study the remains of this natural reactor for the purpose of investigating how nuclear waste decays. What is also interesting is that this reactor appears to have been steered automatically—it never exploded.

I'm not trying to play down the dangers of producing energy from nuclear fission. But we should recognize that hundreds of nuclear power stations are now operating on this planet, and that they have functioned very well for a number of years, with only a few notable exceptions. And let's not forget that some countries are already producing a large proportion of their energy needs in nuclear reactors. Only once, in the mid-1980s, a serious accident occurred in a nuclear power station close to Chernobyl in the Ukraine. In that incident a reactor was totally destroyed and a large amount of radioactive material was released into the atmosphere. Subsequent investigations determined that the catastrophe was due not only to a concentration of unlucky coincidences but also to a remarkable lack of competence on the part of technical personnel. The Soviet government openly admitted as much.

EINSTEIN

Do we have any guarantee that accidents like that won't happen again?

HALLER

We don't. Still, I'm convinced that nuclear reactors can be operated with an acceptable degree of guaranteed security. There is no absolute guarantee.

NEWTON

That I will readily accept. In science and technology there is no room for absolute, apodictic statements. I have experienced this myself these past few days.

EINSTEIN

Let's avoid sophistry here, gentlemen. If things got to the point of a serious nuclear catastrophe in Russia, I'm sure it could happen again. Or do you have reason to believe the contrary?

HALLER

No convincing reason. Still, the Soviet incident has taught us an important lesson, and the mistakes committed there will certainly not be repeated. But there is no guarantee against human error. The engineers responsible for the reactor mismanagement in Chernobyl, for instance, had purposely switched off the automatic security system of the power station in order to do a few experiments, and they were not aware of the dangers involved. The real danger connected with the use of nuclear reactors is related less to technical questions than to political and economic circumstances.

A particular danger lurks if a country that operates nuclear reactors is plunged into armed conflict. There is always the danger of a command raid on a reactor site; an enemy might try to blow up a reactor by means of a small nuclear explosion, for instance. That could make a whole region uninhabitable for a long time. There might be millions of victims. That's a good reason why nuclear reactors should never be built in countries that are politically and economically unstable. But things look different in reality. The political and economic stability of a country or region may change with time. Just think of what happened in the former Soviet Union.

EINSTEIN

Do you think that producing energy in nuclear reactors—in other words, exploiting nuclear fission—is going to solve the energy problem that will certainly pose itself in the future? Even if we can optimistically rule out the political problems you just mentioned?

HALLER

Theoretically, we could easily imagine that energy production by nuclear fission might become the principal source of future energy—and again I'm supposing we don't have to worry about the political aspects. But the way things are developing these days, we could hardly be that optimistic. There are other reasons not to depend on nuclear fission, and one is the question of what to do with the nuclear waste put out by the power stations.

NEWTON

But tell me, is this waste more dangerous than the stuff left over from coal- or oil-burning power plants? I read in a paper in Cambridge that there are plenty of problems with those wastes.

HALLER

Yes and no. It depends on the quantities involved. Burning coal and oil generates toxic materials that penetrate into the atmosphere and do harm to the environment. The waste products of nuclear reactors harm nature in a different way, as a result of their radioactivity. Today we know that most nuclei emitted in fission processes are not stable; in the course of time they decay into different, stable nuclei. In the process, fast-moving particles are emitted, such as the high-energy photons known as gamma quanta.

A lot of radioactive materials occur naturally. The natural radioactivity of our environment must have been much stronger billions of years ago when the Earth was younger; radioactive materials must have been quite prevalent then. But most of the unstable nuclei have since decayed into stable endproducts, so that the elements we find on the Earth's surface today are almost entirely stable. Still, natural radioactivity can't be ignored. We are all exposed to it.

NEWTON

Now I understand what Einstein was getting at in his three-page paper on the energy formula when he spoke of a test of his theory by means of radium salts. Radium must be one of the elements you called unstable, or radioactive.

HALLER

Exactly. The energy radiated by radium (and notice the name of the element!) is strictly nuclear energy. Einstein was entirely correct when he suggested radium as an illustration of the argument in his paper. But in his time he couldn't have known how relevant his formula is in the case of nuclear reactions.

EINSTEIN

I've just recalled a letter I wrote to my friend Conrad Habicht, shortly after my paper on relativity theory was published. I wrote: "A measurable decrease in mass should occur in radium. This idea is both intriguing and amusing; I really can't tell whether the good Lord is having a laugh here and making a fool of me." It now looks as if he wasn't. You see, the Lord is subtle but not malicious.

HALLER

The radioactive particles emitted by unstable nuclei such as radium are very dangerous for biological systems in our environment, and certainly for the human body. They damage biological cell tissue. The fission that occurs in nuclear reactors leaves us with long-lived radioactive substances consisting mostly of heavy elements, and it takes about 10,000 years for the radioactivity of these elements to come down to a level comparable to the radioactivity of the most unstable ores observable in our natural environment.

EINSTEIN

You're telling us we have to store the wastes of a nuclear reactor for at least 10,000 years. That's a long time, and a lot can happen in that time. We certainly wouldn't be happy if our ancestors living in caves 10,000 years ago had left us with their waste products—assuming such long-lived wastes were produced in those days.

HALLER

It's certainly a serious problem. On the other hand, modern technology permits us to deposit radioactive wastes deep in the Earth in locations that are geologically very stable—for instance, in abandoned salt mines a thousand meters or so below the Earth's surface. That guarantees a fair amount of safety. In geologic time, typically measured in millions of years, 10,000 years looks pretty short.

Suppose we compare the radioactivity generated by nuclear fission to the radioactivity naturally occurring on the Earth's surface, which humans have always been exposed to. Suppose, for a minute, we produce as much energy from nuclear fission as can be produced from burning all the known coal deposits in the world—forgetting for the moment that this burning would itself leave tremendous toxic wastes. Even in this extreme case, the radioactivity of all the long-lived radioactive substances produced by nuclear reactors would be negligible compared to the naturally occurring radioactivity—to be precise, it would be only one part in 10,000.

EINSTEIN

That sounds like good news. Still, reactor wastes can't be distributed uniformly over the Earth's crust; there will have to be strong concentrations in a few spots.

HALLER

True. That's why the comparison I just gave you is a bit misleading. But it's useful anyway—it gives us a feeling for the magnitude of the radioactive substances we'd be left with if we were to rely heavily on nuclear reactors. The waste products of these reactors are not a new phenomenon—they simply add to the naturally existing radioactive elements; and this addition is so small that it can, on the average, be ignored.

EINSTEIN

But didn't you just tell us that you don't believe the energy impasse can be solved by nuclear reactors?

HALLER

I think we can reasonably produce energy for a while by nuclear fission as long as the radioactive waste can be controlled. At the present time, to stop producing nuclear energy would be irresponsible. We have no reasonable alternative; burning up valuable raw materials like coal and oil is certainly not a good alternative—not to mention the poisoning of our atmosphere caused by the burning.

In practical terms, I'm thinking of a fifty- to one-hundred-year period. To rely on energy produced by nuclear fission for longer than that doesn't look reasonable to me, unless waste can be kept at a controllable level. That would be possible only if the world population were to be reduced to a fraction of the 5 billion people alive today, maybe to 100 million. But the population is still increasing, and I can't imagine that the energy needed for, let's say, 10 billion people could be generated by fission reactors without the danger of occasional catastrophes and radioactive poisoning of large regions.

EINSTEIN

In other words, nuclear fission doesn't seem to offer a real solution to the energy problem.

HALLER

No, it doesn't. Anybody claiming to have a perfect solution lacks credibility. In the future, the problem of energy production will be solved only slowly and painfully. A number of possibilities will have to be explored, and nuclear energy will be only one of them. There is also the need to conserve energy through the many techniques available in our modern world. We should remember that energy we save, by electronics, for example, is energy we don't have to produce in the first place.

In southern countries, energy—particularly electrical energy—will be generated more and more by exploiting the sun. Also, let's hope that in the long run there will be a slow but steady reduction in world population. As far as a solution for the energy problems of the future is concerned, you can see that I'm neither euphoric nor particularly pessimistic.

NEWTON

I notice that when you talk about energy production you don't mention nuclear fusion.

HALLER

I did tell you yesterday that it's still not clear whether we will ever be able to produce energy from fusion—and I'm speaking only of things we can do on Earth now. In the sun, fusion is occurring all the time, and the energy we can produce from the sun's rays that reach Earth might be called indirect fusion energy. That may be as far as we can get with nuclear fusion. But even if we were to succeed in controlling fusion, we still couldn't be sure that energy could be produced in technologically useful ways. One thing is sure: we have a long way to go from today's research to possible applications.

EINSTEIN

And how about the wastes from fusion?

HALLER

Like all nuclear processes, nuclear fusion will leave radioactive wastes. But here fusion has the advantage: the radioactive elements are mostly light ones and not very long-lived; they certainly don't present a 10,000-year danger. Unfortunately, heavy elements such as metals have to be used in constructing a fusion reactor. Since these will be made radioactive by the fusion process, a fusion reactor might also give us a problem with radioactive wastes, but I don't think it would be nearly as serious as in the case of fission reactors.

I don't want to seem overpessimistic about nuclear fusion. It's just a matter of our not knowing today whether anybody will be able to build a fusion reactor in a way that results in an economical energy gain. No doubt, research should be encouraged, but you know that success in scientific research can't be programmed. Success may appear out of the blue, or we may have to wait a very long time for it. Sometimes there is no success because the method employed turns out to be impractical. But even if we were to find out that nuclear fusion permitted an economically feasible method of energy production, maybe fifty years from now, that hardly means we would have as much energy as we pleased. For one thing, there would still be a waste problem, and for another, it might not be a good thing to have that much energy. Experience shows that when energy is plentiful, our society wastes it recklessly. And that's one thing we shouldn't wish on ourselves if we want to protect our environment.

This may sound illogical, but I believe firmly that human civilization can continue on our planet only if we manage to use our energy sources and our raw materials very sparingly, and that includes nuclear energy. I believe this is our only chance.

I looked at my watch—the morning had almost gone. A scientific colleague from the United States, an old friend of mine, was walking in; he nodded at me and looked curiously at our little group. We greeted each other, but I was careful not to introduce my companions with their full names, mentioning only their first names. My friend agreed to give my two companions a tour of CERN that day. We would meet again late in the afternoon in the office up in the theory division that I use whenever I spend a few days at CERN.

NINETEEN

Mysterious Antimatter

Sitting in my office, I heard Einstein's and Newton's voices in lively discussion as they walked up the corridor to my office door. It was clear that the afternoon's laboratory visit had impressed them deeply.

For a while we discussed the big accelerators at CERN, known by their acronyms SPS (Super Proton Synchrotron) and LEP (Large Electron Positron). As its name indicates, the first of the two machines accelerates protons. Their total energy can be raised to as much as 400 GeV, which is more than 400 times the energy that corresponds, according to Einstein's formula, to the rest mass of the proton. The same machine also accelerates the protons' antiparticles, commonly called antiprotons, at the same time as it accelerates protons. (We will deal with antiparticles and, more generally, the concept of antimatter a little further on.) In this manner, protons and antiprotons can be made to collide head-on, and the particles generated in these violent collisions can be analyzed in special detection devices. These particle detectors measure the trajectories of the particles in a number of different ways.

The other machine, LEP, accelerates both electrons and their antiparticles, commonly called positrons, to energies of about 50 GeV each. Here again, particle detectors surround the interaction points to study the head-on collisions that occur in LEP. Like the accelerator itself, these massive detectors are installed in underground tunnels and halls.

NEWTON

Your friend, our guide, was kind enough to explain the essential features of the accelerator to us. When he went on to talk about the research projects for which these machines were built, he kept referring to "antiparticles" and "antimatter." The fellow obviously didn't have the foggiest notion to whom he was talking; and so he assumed that a couple of theorists like us would know all about antimatter and the like, and didn't offer a word of explanation. You will understand that we didn't ask any ques-

Fig. 19.1 One of the large detectors used at CERN to study proton-antiproton collisions at high energies. (Courtesy of CERN.)

tions of him, so I shall have to ask you: What is an antiparticle and what is antimatter?

HALLER

To answer those questions, I'll begin by going back to Cambridge. In the late 1920s a young physicist by the name of Paul A. M. Dirac established a link between atomic physics, which was then rapidly progressing, along with its underlying theory, quantum mechanics, on the one hand, and Einstein's theory of relativity on the other. By the way, Dirac later occupied the same chair at Trinity College that was once yours, Professor Newton.

It soon became apparent that linking these various fields was not that easy; but for most practical purposes in atomic physics there is no need to pay attention to relativistic effects anyway. Typically, the speeds of particles inside atoms are a good deal smaller than the speed of light, so space and time play quite different roles in atomic physics, essentially as in classical mechanics. Dirac was not so much in search of a novel explanation of certain phenomena in atomic physics as attempting to pursue his ideas

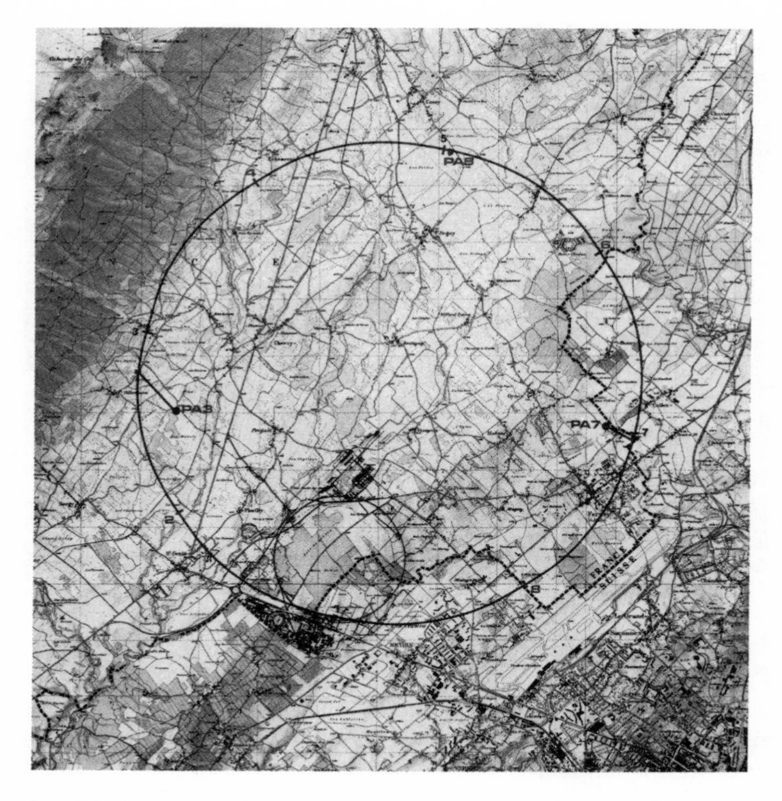

Fig. 19.2 The LEP accelerator is housed in a 27-kilometer-long ring-shaped tunnel running below ground level in the sloping valley between Geneva airport and the Jura mountains. The indications along the ring mark the interaction zones where collisions between electrons and positrons can occur. The smaller ring indicates the tunnel of the SPS accelerator; the densely built-up area below is the laboratory site proper. The LEP machine has been in operation since 1989. (Courtesy of CERN.)

as a matter of principle; he wanted to express the new insights of atomic physics in such a way that they could be generalized in terms of the theory of relativity.

EINSTEIN

In relativity theory there is really no room for a basic difference between space and time. They are inextricably intertwined. I presume Dirac set out to work out the details of this linkage in the case of quantum mechanics.

Fig. 19.3 Artist's view of one of the LEP experimental zones. Electron-positron collisions occur deep underground, so the massive particle detectors have to be housed in subterranean halls.

HALLER

That's right. Dirac simply started from the assumption that in all cases dealing with particles traveling close to the speed of light, space and time have to be treated in the same manner. That's how he managed to derive, in 1928, a very interesting equation describing what is now known as the Dirac function, which ensured an extension of relativity theory into atomic physics. Its first application achieved instant and surprising success: it described quite precisely the interactions of electrons with magnetic fields.

Soon afterward, Dirac noticed that his equation was not only capable of correctly describing the behavior of electrons in the presence of atomic forces; it also predicted the existence of a new set of particles. These new particles would have the same mass as electrons but the opposite electrical charge—a positive charge of the same magnitude as the electron's negative charge.

Dirac himself was unwilling at first to accept the full consequences of his theory, since the only positively charged components of atoms are protons, which have masses much larger than those of electrons. After many unsuccessful attempts at finding another solution to this problem, Dirac finally convinced himself that these additional particles should actually exist alongside electrons and protons. He called them the antiparticles of the electrons.

An important aspect of Dirac's theory of electrons is that electrons and their antiparticles show up in a completely symmetric fashion. They can be exchanged for one another without any essential change becoming apparent.

But now let's leave Europe and go to the California Institute of Technology in Pasadena. At Caltech, in the early 1930s, detailed studies in cosmic-ray physics were being pursued. For this purpose, one of the researchers, Carl Anderson, built a cloud chamber, which was quite large by the standards of that time and could be used to observe and photograph the tracks that electrically charged particles leave in the chamber after traveling through it. If a chamber of this kind is brought into a strong magnetic field, the particles will travel through it on curved paths. The amount of curvature and its direction can be used to determine the mass as well as the charge of a particle. Anderson observed a large number of such paths, and convinced himself that all the tracks were due to known particles—either electrically positive protons or negatively charged electrons.

In the morning of August 2, 1932, Anderson was studying the photographs of the latest cloud chamber tracks, as he did every morning. But on that day he noticed a track that at first glance appeared to be that of an electron—its mass turned out to be equal to the electron mass. But the curvature was wrong. It was in exactly the opposite direction to what would be expected of an electron. It behaved like a positively charged electron.

Anderson's natural reaction was to check his apparatus to exclude all possible experimental mistakes. He didn't find any. In due course he found more of the mysterious particles. Some of their tracks appeared to stop abruptly, as though the particle had suddenly vanished. There was no longer any room for doubt: the electron had an electrically positive twin brother. Anderson called his new particle a positron.

EINSTEIN

In other words, Anderson discovered the particles that Dirac had predicted.

HALLER

Yes and no. At the time he was carrying out his experiments, Anderson had no knowledge of theoretical work that Dirac had done in Cambridge. It was only in 1933 that the particles Anderson had discovered were recognized as the antiparticles whose

Fig. 19.4 Carl Anderson standing by the cloud chamber that enabled him to discover the positron. The chamber proper is hidden inside the black coil, which generated the magnetic field for the experiment. The inset shows one of the first positron tracks found by Anderson. (Courtesy of California Institute of Technology.)

existence had been postulated by Dirac. The positron was then firmly established as the antiparticle of the electron, and given the symbol e^+. As I said, its mass is identical to that of the electron:

$$m(e^+) = m(e^-) = 9.1091 \times 10^{-31}\,\text{kg} = 0.511\,\text{MeV}.$$

NEWTON

Does the proton also have an antiparticle?

HALLER

Anderson's discovery was really the first step into a completely new realm, the realm of antimatter. Today we know that every particle has its antiparticle. Neutral particles may sometimes be identical to their antiparticles; the photon, for instance, is a hybrid particle—it is simultaneously its own antiparticle.

So the proton also has an antiparticle, denoted by the symbol $\bar{p}$. The antiproton has the same mass as the proton but a negative electric charge. It was discovered in 1955—not in cosmic radia-

Fig. 19.5 The CERN antiproton accumulator uses magnetic fields to contain antiprotons in its ring-shaped vacuum chamber before injecting into the SPS accelerator.

tion, like the positron, but in an accelerator experiment at Berkeley. Still later, the antiparticles of neutrons were discovered. Like neutrons, antineutrons are electrically neutral—they carry no electric charge.

Here at CERN, techniques have been developed for the production of large numbers of antiprotons. It's done by means of a small accelerator that feeds them into the large SPS accelerator, which subsequently raises their energy, just as it does with protons, until they are moving at a speed close to the speed of light.

EINSTEIN

If I had a positron and an antiproton here, wouldn't they attract each other due to their opposite electric charges, just like an electron and a proton? And if so, doesn't that mean I could construct something like an atom out of these antiparticles?

HALLER

Of course. We could produce a new element that way, antihydrogen. If we included antineutrons, we could also—theoretically—produce more complicated antielements such as antihelium or anti-iron, even antiuranium. But only, as I said, theoretically; experimentally, no one has managed to produce antinuclei heavier than antideuterons and antihelium. Still, we should be aware that all matter in our universe has its antimatter in principle, if not in reality. There is no basic difference between matter and antimatter—this is one of the important things we learn from Dirac's theory.

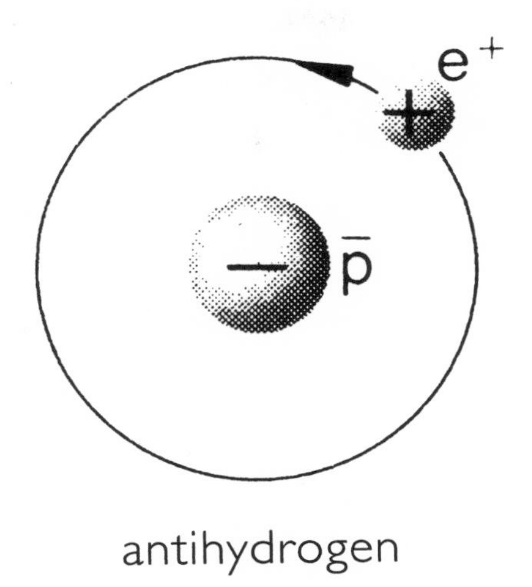

Fig. 19.6 The simplest form of antimatter is antihydrogen. It consists of an antiproton and a positron.

NEWTON

Didn't you say earlier that some of the positron tracks in the cloud chamber stopped all of a sudden? What happened there? Did the particles vanish?

EINSTEIN

Let me ask another question about those tracks: Where did they start? Where did the positrons come from, in Anderson's experiment? If they can vanish, they could also suddenly appear.

HALLER

You're absolutely right. Those two phenomena are connected, as we'll see in a minute. But first about those tracks that quit: they too can be explained by Dirac's theory. As I mentioned, this theory provides a kind of synthesis between relativity and atomic physics, so it's no surprise that Einstein's formula comes to bear here again. And it does so in a very impressive fashion.

Let's take a look at a positron and an electron colliding with each other. At the instant of the collision a catastrophe occurs in this microphysical world—the two particles annihilate each other. What remains is energy in the form of electromagnetic radiation, that is, photons. The mass of the two particles has been transformed during this annihilation process into the energy of the photons, strictly according to Einstein's rule of the equivalence of mass and energy.

NEWTON

That tells us what Anderson saw. Whenever he saw a positron track that had suddenly stopped, the particular positron had collided with an electron, and the two had annihilated each other.

HALLER

That's exactly what happened. Dirac's theory even prescribes how to calculate the number of photons generated in the annihilation process: there will usually be two. What Anderson saw most of the time was merely the disappearance of the positron. His cloud chamber didn't permit him to observe the two photons leaving the annihilation point at the speed of light—electrically neutral particles such as photons don't leave tracks in the chamber.

NEWTON

There you have it, Einstein—this extraordinary transformation of particle and antiparticle into nothing but photons is the most extreme consequence of your formula. The entire mass turns into energy, not just a small fraction as in nuclear fission or nuclear fusion. This is what I've been waiting for! Finally, here is a process that converts an entire mass into radiation energy.

EINSTEIN

Since we know the respective masses of an electron and a positron, it's easy to determine the energy of each of the two photons. If the two particles meet in slow motion so that we can ignore their kinetic energy, it's only the rest mass of the particles that is being converted. But the rest mass of the electron is 0.511 MeV, so we can predict that the energy of the two photons leaving the annihilation point must be about 0.511 MeV each, provided we look at the process from a reference system in which the annihilation takes place at rest.

HALLER

And that's exactly what was found; theoretical expectation and experimental measurement agreed amazingly well. In this respect, the annihilation of electrons and positrons is the most impressive

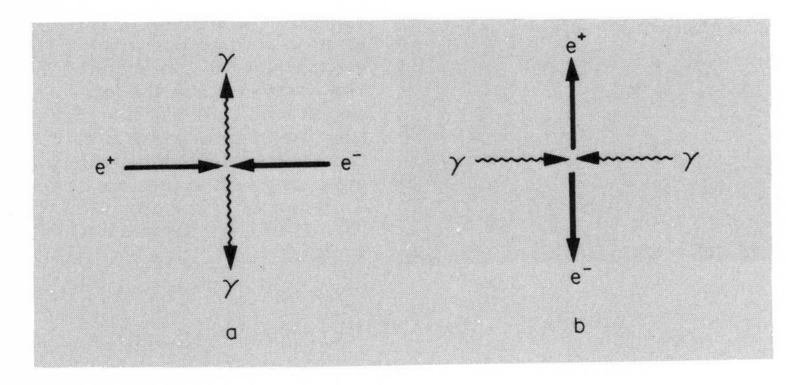

Fig. 19.7 (a) The annihilation of an electron and its antiparticle, the positron, leads to the emission of two photons (or gamma quanta). The energy of the annihilating particles equals that of the emitted photons. (b) Conversely, two colliding photons can generate an electron-positron pair. According to Einstein's equation, this process can occur only if the available energy is sufficient for the creation of the two massive particles.

illustration of your formula and, less directly, of the theory of relativity.

EINSTEIN

But you still haven't told us where positrons come from.

HALLER

Not directly, but I've suggested it indirectly. We've just seen how an electron and a positron can annihilate themselves into two photons. At this point it's worth recalling a basic law of physics that applies in Dirac's theory as well as in Newton's mechanics and Einstein's relativity theory: every microscopic process is reversible. It's often called the law of time reversal invariance. We've already said that when electron and positron meet head-on, they convert all their mass and energy into two photons. Let's turn this around and have two photons interact with each other: the interaction may result in their converting into an electron-positron pair. Note that these two particles didn't exist previously. They are being created from "nothing"—or, to be more precise, from energy. This creative process is the exact inverse of electron-positron annihilation. So the creation and the annihilation of particles are intimately related processes.

Fig. 19.8 The creation of an electron-positron pair in an electromagnetic interaction. The arrow indicates the location at which the pair materializes. The opposite curvatures of the tracks in the magnetic field are clearly visible; they are a consequence of the opposite charges of electrons and positrons.

EINSTEIN

That's another version of the old principle: no production without destruction, no birth without death. But keep in mind that the particle production you mentioned can happen only if the energy of the two photons is sufficient to generate at least the rest masses of the particles, according to my formula.

HALLER

Of course, the two photons must have a fair amount of energy to provide for the mass of an electron-positron pair, that is, at least twice 0.511 MeV. If they don't, nothing will happen at all, and the photons will just pass each other without interacting.

NEWTON

Let me ask another question. The electron and the positron, by definition, have opposite but equal electric charges. In this respect they behave just like an electron and a proton. But if I bring an electron in contact with a proton, I produce a hydrogen atom. Doesn't it occur to you that something similar might happen with an electron-positron pair? In other words, isn't there a kind of atomic structure just before the two particles' annihilation?

HALLER

Sir Isaac, you've gotten ahead of me again. This atomlike configuration does exist. It's called positronium. Whenever an electron and a positron approach each other in slow motion, they wind up as positronium. It's not really an atom; unlike a hydrogen atom, it has no nucleus. Both electron and positron have the same mass, so the two particles orbit around each other. And positronium is very short-lived: less than one-millionth of a second after its creation it decays into photons—typically into two photons.

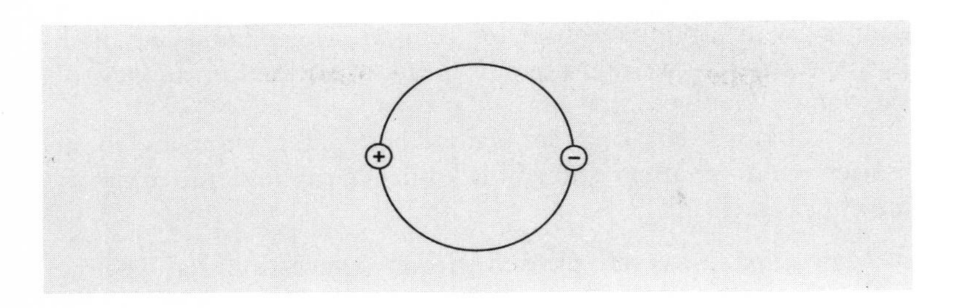

Fig. 19.9 Positronium is a configuration of matter and antimatter (more precisely, of an electron and a positron) that resembles a hydrogen atom. Both particles move in circular orbits. Less than one-millionth of a second after its creation, the positronium "atom" decays into two or more photons.

EINSTEIN

All these new insights boggle the mind. To think that physics has come up with all this since I published my first paper on relativity theory! This positronium business is pretty odd. Just imagine—matter and antimatter squeezed together in a tiny space, and before they convert into pure radiation, they still have time to make up a kind of atom on the way.

As for antimatter, so far we've discussed only individual antiparticles that happen to be produced in particle interactions. But maybe we should think of larger amounts of antimatter somewhere in our universe, say, antihydrogen or anti-iron. If a sizable chunk of anti-iron somehow made it to the Earth's surface, it would immediately annihilate the ordinary matter all around it. The result would be a gigantic explosion, far more powerful than a hydrogen bomb detonating.

HALLER

That would depend on how quickly matter and antimatter were allowed to approach each other. Certainly, if we put one kilogram of antimatter into a pot containing the same amount of normal matter, it wouldn't take much stirring to have all hell break loose. But we might try to have the annihilation occur slowly and even use it to create energy.

Technically, this wouldn't be difficult, because unlike nuclear fusion the annihilation of matter and antimatter doesn't need high temperatures; the process occurs fairly easily. It would take only one kilogram each of antimatter and matter, slowly annihilating each other, to provide the energy needed for an entire year by a country with a population of ten million. There's just one hitch in this scenario—we don't own a chunk of antimatter. In view of the danger involved, that may be just as well.

But look, it's dark outside. We had better start worrying about matter again—matter of an edible kind. We can revert to antimatter after that.

My companions were pleased with my suggestion that we dine in a restaurant across the border, in France. We got into my car and headed for the Jura mountains.

TWENTY

Marveling at Elementary Particles

It took us only twenty minutes to reach our destination. La Fortune du Pot is a small, comfortable restaurant nestled just below the Jura mountains in the village of Saint-Jean de Gonville. The innkeeper seated us at a corner table, where we would be undisturbed. After Einstein had studied the wine list, we followed his advice and ordered a bottle of Châteauneuf-du-Pape. From the menu we chose a venison dish that I had had before and could warmly recommend. Putting down the wine list, Einstein returned immediately to antimatter.

EINSTEIN

More than once you've spoken of symmetry between matter and antimatter. But when I look at the universe, I see no such symmetry. All I see around us is matter—I see the three of us, this table, the air we breathe. Even the venison we're waiting for is more likely to have started out as a deer rather than an antideer. We live in a world of matter. Why don't we see antimatter in the cosmos? Or are you telling me that somewhere out in space there are stars and planets made of antimatter?

HALLER

You have brought up an important issue. I'm afraid we don't have a satisfactory answer to your question even today. But I can cite a few facts. A star consisting of antimatter would look just like a normal star. Its light would be exactly the same, whether it came from particles of matter or antimatter.

NEWTON

Then there might be some antistars in the Milky Way? Could it be that our galaxy is half matter and half antimatter? On the average, there would be just as much matter as antimatter, and the symmetry between matter and antimatter of which you spoke earlier would be realized at least in the mean. By sheer coincidence we would be on a matter planet; some other civilization might be housed on an antimatter planet.

HALLER

But that's not actually the case. If there were an antistar somewhere in the Milky Way, we would certainly witness frequent occurrences of annihilation because the antistar could not avoid getting in contact with normal matter in its proximity. As a result, many energetic photons would be emitted. Gamma-ray sources of this kind have been searched for, but without success. So we're quite certain today that in our galaxy, at least, there is no antimatter beyond the few antiparticles produced in clashes of normal particles, such as the positrons discovered by Anderson.

Only a few kilometers from here, at CERN, they produce a great quantity of antiprotons. Without exaggerating, we can say that CERN, and a similar lab near Chicago, the Fermi National Accelerator Laboratory, are the only places in our galaxy where you will find antiprotons in fairly large quantities. But even there we're dealing with minuscule, submacroscopic amounts.

Even an amount as small as one gram of antimatter doesn't exist in concentrated form anywhere in our galaxy. And we should be happy that there are no antistars in our galaxy. If nature had followed your advice, Professor Newton, and had constructed the Milky Way out of stars and antistars in a fifty-fifty mix, we probably wouldn't exist at all. The Earth would continuously be bombarded by energetic protons emitted from annihilation processes; that would have catastrophic consequences for life on our planet. It's doubtful that life could have developed at all.

When it became clear during the twentieth century that the Milky Way consists only of matter, it was natural to suggest that there might be other galaxies in space consisting only of antimatter. But by now we are sure that this is not so. There are processes that lead to the exchange of small amounts of matter between galaxies, but in known cases of contact, no annihilation processes have been observed. That most probably means that the universe, as far as we can see it through our telescopes today, consists only of matter. Nature appears to have discriminated against antimatter. In our physics labs, antimatter particles may well appear symmetrically with matter particles, but in the architecture of the universe they seem to play no role.

Fig. 20.1 A distant galaxy that sometime in the past collided with another, smaller galaxy. The process involved an intensive exchange of matter, which did not radiate off, suggesting that both galaxies consisted of matter and that even distant galaxies are made of matter, not antimatter.

NEWTON

Yet there must be hypotheses that attempt to explain this strange phenomenon. If I were active in physical research today, a problem like that would be just to my taste.

HALLER

There are plenty of hypotheses. One interesting theory, which probably has some truth to it, starts from the so-called Big Bang around 15 billion years ago. According to this theory, matter and antimatter existed in symmetry in the beginning. The symmetry was imperfect, however; there was a very small excess of matter over antimatter particles—a single matter particle for every 10 billion pairs of matter and antimatter particles. These pairs annihilated each other in the course of time, and in the end only the excess matter particles remained. Those are the ones from which our world, and we ourselves, have arisen.

Our conversation was interrupted as the food was being served. For a while we devoted all our attention to the excellent meal. Then Einstein resumed the discussion.

EINSTEIN

You mentioned this strange primeval bang earlier, Haller, but I admit I didn't really understand the business of the missing antimatter. For one thing, why was there an excess of matter over antimatter in the first place? What was the Big Bang anyway? Are we sure there was a primeval explosion?

HALLER

We shouldn't stray too far from our main topic—remember, we meant to concentrate on relativity theory and things closely connected with it. If we get into cosmology, we'll soon forget our original topic. We could spend days, maybe weeks, discussing complicated issues of astrophysics, particle physics, and cosmology.

EINSTEIN

You're probably right. We'll postpone the Big Bang discussion to a later date.

NEWTON

I agree. Even though the most recent hypotheses about the origin of the world are of great interest to me, so are the hard facts concerning antimatter. I must also bear in mind that I have to be back in Cambridge tomorrow evening. Let's get back to antimatter, then. So far, we have considered only the annihilation of electrons and positrons. When an electron collides with a positron, they destroy each other, leaving two photons.

HALLER

That's true only when the two particles move relatively slowly. When electrons and positrons collide at speeds close to that of light, different processes result. Experiments of this kind have been performed in several places, for instance, in a large German national laboratory called DESY (German Electron Synchrotron) in Hamburg.

If we make electrons and positrons collide head-on at energies of many GeV, annihilation often takes the form of a microscopic fireball. Like a phoenix from the ashes, a whole slew of elementary particles arises, including protons and antiprotons. One condition that must be satisfied is that the sum of the energy of all particles created has to equal that of the original electrons and positrons. The total energy remains constant.

There are even reactions that produce very heavy particles. Let me give you an instructive example. If we shoot an electron and a positron at each other in such a way that each has an energy of 4.7 GeV, they will meet almost at the speed of light. In the collision, a new particle may be produced whose mass corresponds exactly to the total energy of the colliding particles, that is, 9.4 GeV. A particle of this kind, about ten times as heavy as a proton, was discovered in the United States in 1977 in a different reaction—but it was observed in electron-positron annihilations at the DESY lab about a year later. It's now called an upsilon, or "Υ particle."

NEWTON

The creation of such a massive particle should be seen as a spectacular verification of Einstein's equation, since all the motion energy of the electron and the positron is converted into the mass of the newly created particle.

EINSTEIN

And what happens to this particle once it's been produced? Does it just sit there?

HALLER

Not at all. It lives for a very short time and then decays into other particles; it may even turn back into an electron and a positron. But its decay can also leave us with a proton and an antiproton, a neutron and an antineutron, or a variety of other particles.

NEWTON

What's the significance of these superheavy particles that live for such a short time? What do we know about Υ particles, for example?

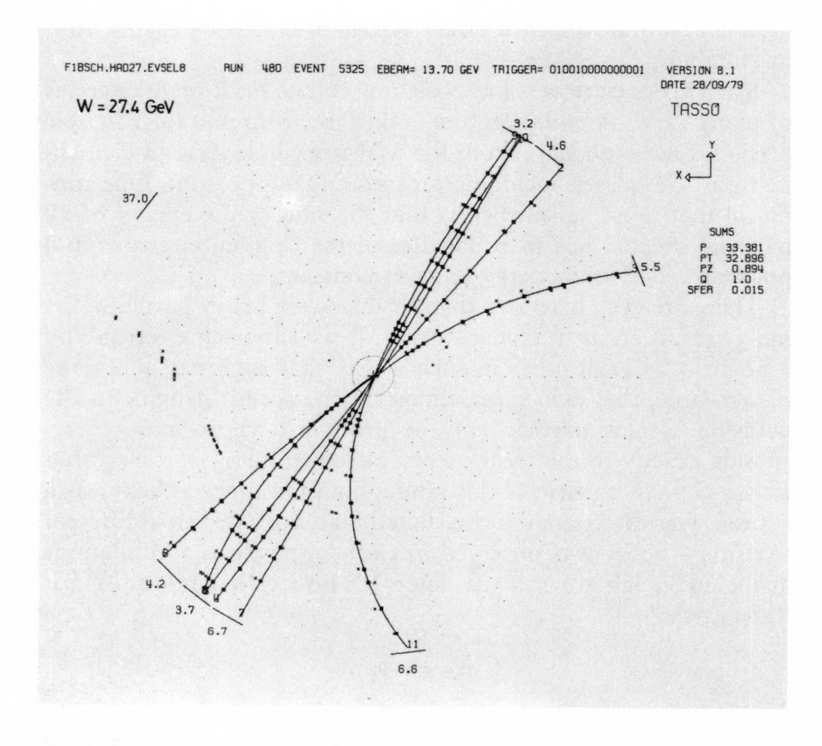

Fig. 20.2 An electron-positron annihilation event in which the total energy of the colliding particles amounted to 27.4 GeV. In the process, eleven charged particles were created. The event was registered by the TASSO Collaboration at the DESY Laboratory in Hamburg, Germany, in 1979.

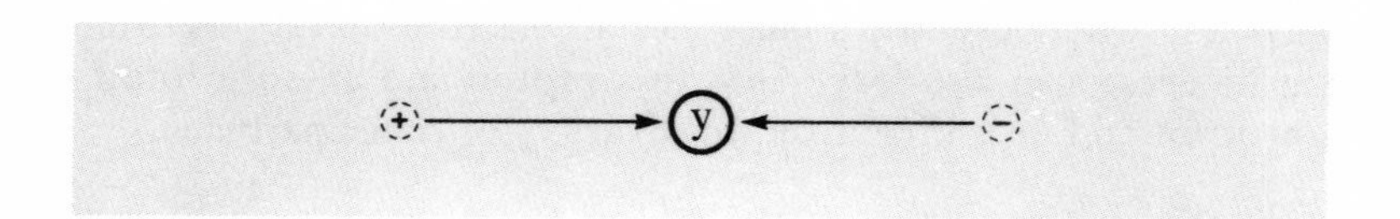

Fig. 20.3 An electron and a positron, both with 4.7 GeV of energy, collide, forming a heavy upsilon meson in the process. Since the mass of this meson equals the total energy of the colliding electrons, creation has occurred at rest.

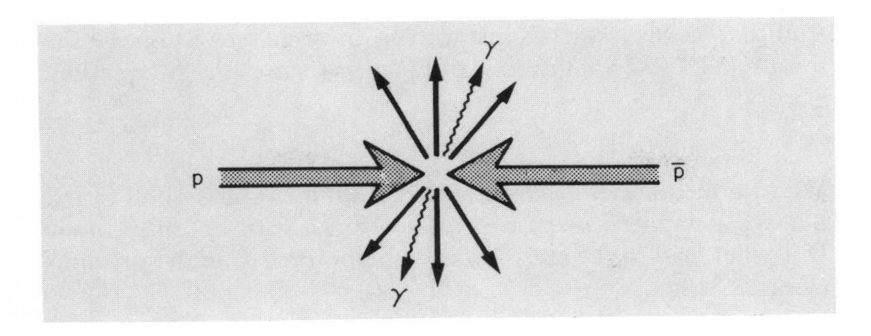

Fig. 20.4 A proton-antiproton annihilation event, with the subsequent emission of many particles, mostly mesons—unlike an electron-positron annihilation, which frequently results in the emission of only two photons.

Fig. 20.5 A bubble chamber showing the annihilation of antiprotons upon impact on atomic nuclei in the chamber fluid. The three curved tracks entering the chamber from the left are due to relatively slow antiprotons. Each of these will ultimately hit a nucleus—in the present case, a proton. In the subsequent annihilation process, several mesons are created. The central antiproton track is clearly seen to lead to an annihilation vertex from which four charged meson tracks issue. (Courtesy of CERN.)

HALLER

Your question leads to the same dilemma as the cosmology we touched on earlier. If we should get into the details of particle physics, it would take days just to get a halfway satisfactory overview. My example was simply meant to illustrate that modern particle accelerators can be used to produce very heavy particles from the collision of such lightweights as electrons and positrons.

And that, as Sir Isaac has just told us, must be seen as one of the most remarkable applications of Einstein's mass-energy equation.

NEWTON

We've talked about electron-positron annihilation several times. But what happens when a proton collides with an antiproton? Does that lead to the emission of two photons, as in the previous case?

HALLER

It could happen. If both these particles were almost at rest before the annihilation, the total disposable energy would simply be double the mass of the proton, that is, about 1.88 GeV. In principle, this energy could be rediscovered in the energy of two photons leaving the annihilation region.

But when the experiment is actually performed, researchers usually find very different processes occurring. A whole series of particles is produced, including photons as well as electrically charged and neutral massive particles. I really can't go into the details of this phenomenon—it has to do with the structure of the colliding particles. Protons, after all, have an internal structure that is very different from that of electrons.

EINSTEIN

But what kind of particles do they find? Electrons and positrons?

HALLER

They often find particles we haven't even talked about, called π mesons, or pions. But before I discuss these particles I should mention that they are frequently produced when protons or even nuclei collide with each other. Take a look at this photo of a particularly energetic collision.

Einstein and Newton looked with great interest at the photo I had placed on the table (see fig. 20.6). It shows the final state of an interaction in which highly accelerated nuclei of the element sulfur were shot at gold nuclei. The sulfur had a total energy of 6,400 GeV, and the gold was at rest.

Fig. 20.6 At the far left a highly relativistic sulfur nucleus has collided with a gold nucleus at rest. The energy of the sulfur projectile was 6,400 GeV. Hundreds of new particles are created in the process, while both the initial nuclei are shattered. The highly specialized detector that took the photographic record is called a streamer chamber. (Courtesy of CERN.)

EINSTEIN

My goodness, hundreds of particles are resulting from the interaction.

HALLER

Many of the tracks you see here are simply the protons that were originally part of the sulfur nucleus, which came in from the left. But most of the tracks were made by the mesons that were produced in the collision.

NEWTON

And once again they accord with your formula, Einstein. It looks to me as though these mesons come pretty cheaply. They are obviously produced readily and in large quantities, when protons or atomic nuclei collide. We just have to make sure that there is enough energy available to keep Einstein's formula fulfilled.

EINSTEIN
Please, Haller, tell us more about this odd new breed of particles.

But at that moment the innkeeper's wife came in with the desserts. Before long, we left the hospitable place. There was a full moon, and the Jura mountains reflected its glow. Einstein wanted to take a few steps through the streets of Saint-Jean de Gonville and the meadows beyond. By the time we got back to the CERN hostel, it was almost midnight.

TWENTY-ONE

Does Matter Decay?

Next morning I joined Einstein and Newton for breakfast in the CERN cafeteria before we retired to my office. There was not much time left for discussion; part of that day, the last we would share, was set aside for Newton and Einstein to observe some of the major experiments being performed in the laboratory.

NEWTON

We just touched on mesons yesterday. Why don't you tell us a little more about those strange objects?

HALLER

You recall our talking about muons, the particles produced in the upper atmosphere that take a relativistic dive to the surface of the Earth?

EINSTEIN

Yes, but that works only because of time dilation.

NEWTON

How could I forget those particles? They are to blame for hastening the collapse of my old edifice of space and time.

HALLER

When I told you that muons are generated in the upper atmosphere, in the collisions of cosmic-ray protons with the atomic nuclei of the atmosphere, I was being slightly inaccurate. What these collisions first produce are mesons—more precisely, π-mesons—which travel through space for a very brief time and decay almost immediately into muons and neutrinos.

EINSTEIN

Fine, but what is the nature of these mesons? Are they related to protons?

HALLER

Suffice it to say that there are three kinds of π-mesons distinguished by their electric charges—positive, negative, and neutral. Their masses are almost equal—about 140 MeV, which makes them about 30% heavier than muons. They are denoted by the symbols π^+, π^-, and π^0. Muons are generated by the decay of the charged mesons, π^+ and π^-, which have an extremely brief lifetime—on the order of 10^{-8} seconds, one-hundredth of a microsecond.

NEWTON

And how do the neutral mesons decay?

HALLER

Neutral mesons are much shorter-lived than the charged ones; they live for only 10^{-16} seconds. It's therefore extremely difficult to determine their lifetimes experimentally. For all practical purposes, neutral mesons decay into two photons immediately after their creation.

EINSTEIN

As with positronium decay, where electron and positron annihilate each other.

HALLER

That's a very relevant observation, Einstein. It has been noted that a meson is in fact very similar to positronium, which is a matter-antimatter object.

NEWTON

Are you going to tell me that mesons consist of electrons and positrons? If so, what do the electrically charged ones look like?

HALLER

That's not exactly what I meant. But mesons do consist of matter and antimatter. Today, we know that protons, neutrons, and hence all atomic nuclei are actually composites of even smaller particles called quarks. A proton, for instance, consists of three quarks, and an antiproton consists of three antiquarks.

EINSTEIN

How strange! When we were looking at pictures of high-energy particle collisions yesterday, we saw numerous tracks. But you told us that all of them were either protons or mesons. How come there were no quarks among them? If the collisions occurred at such high speeds, I would suppose they would have shaken off some quarks from the nuclei.

HALLER

To this day, no one has observed quarks directly as isolated particles in the laboratory. Quarks display dual behavior: inside atomic nuclei they behave as normal particles—and in this respect they are observable, if only indirectly. But when we attempt to remove one quark from its fellows, we always fail. The farther we remove one quark from another, the stronger the forces that bind them together. Consequently, we are pretty sure that it will never be possible to observe quarks as isolated particles. When I spoke of quarks as objects inside protons, I purposely did not call them particles.

NEWTON

Now I know what you meant when you said mesons are really a state of matter and antimatter. You mean they consist of a quark and an antiquark?

HALLER

Precisely. In this sense a meson can be compared to positronium, which is also a combination of particle and antiparticle. But back to quarks. Current ideas about the nature of all matter are based on the existence of quarks. And I have another surprise for you: there are several kinds of quarks, and charged mesons consist of a quark of one kind and an antiquark of another. That makes it possible for some mesons to have an electric charge. Different kinds of quarks have different charges; a meson's charge is simply the sum of the charges of the quarks inside it.

The neutral π-meson consists of a quark-antiquark pair of the same kind. That makes it look very much like positronium. If we replace an electron and a positron by a quark and an antiquark of the same kind, we get a neutral meson. That explains the short lifetime of the meson—quark and antiquark like nothing better

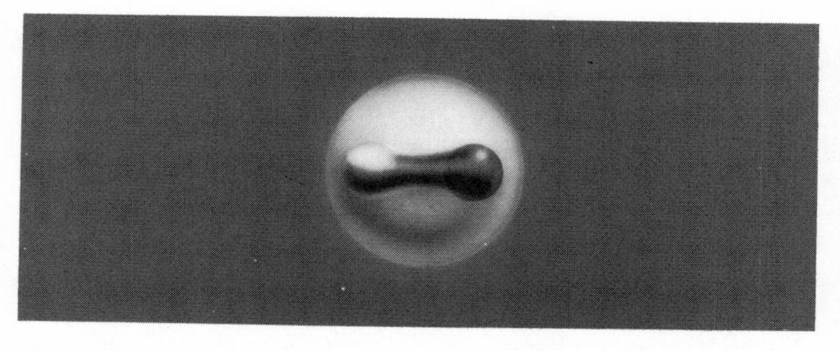

Fig. 21.1 The inner structure of a meson, showing its composition of a quark and an antiquark. Some electrically neutral mesons decay by means of quark-antiquark annihilation into photons, analogous to electron-positron annihilation of positronium.

than chasing and annihilating each other, and this is what happens immediately after the creation of the meson.

EINSTEIN

So this is the special feature of mesons. They are matter and antimatter bound together, a bound state of energy, and this energy can be released by the decay of the particle right after its birth.

HALLER

That's one way to put it. It makes neutral mesons short-lived witnesses of your formula. At their decay their entire mass of about 2×10^{-25} grams is converted into radiation energy.

NEWTON

We've talked so much about unstable particles that I'm afraid protons and atomic nuclei might turn out to be unstable too. How do we know that protons *are* stable? Isn't it odd that in elementary particle physics we seem to be dealing with unstable particles all the time, yet atomic nuclei are stable?

HALLER

That's a very important point. Today we believe that the matter in and around us was generated about 15 billion years ago in the Big Bang. There appears to be a strict rule in nature: whatever can be produced can also decay. Birth and death are intimately connected. If nuclear matter was generated at some point, it can also decay. It's therefore assumed that even protons are ultimately unstable and will also decay in the course of time.

EINSTEIN

And how could a proton decay?

HALLER

One interesting possibility, in fact the simplest one, is by decaying into a positron—which to a certain extent takes over the proton's electrical charge—and a neutral meson. The meson, of course, would immediately decay into two photons.

NEWTON

A positron emerging from the decay of a proton! Take hydrogen, then: when its proton decays and emits a positron, this positron could then collide with the electron in the shell of the hydrogen atom. And that would result in the annihilation of the electron and the positron into two photons. Now look at this, Einstein: the whole hydrogen atom dissolves into four photons—pure radiation energy. Another application of your formula!

EINSTEIN

I'm sorry, Haller, but I really can't accept that. If the proton is bound to decay, we have a much more serious question to answer: Why are there any protons left in the world? Why haven't they all disintegrated? Why do we exist here?

HALLER

I can think of only one answer: the proton has a very long lifetime. There are some fairly specific theories that suggest that the average proton exists for about 10^{33} years—about a million billion billion billion years.

NEWTON

That defies the imagination. Does it mean that a proton's decay can never be observed?

HALLER

No, that doesn't follow. The figure I mentioned is only the average lifetime of a proton. If you want to observe proton decay, even

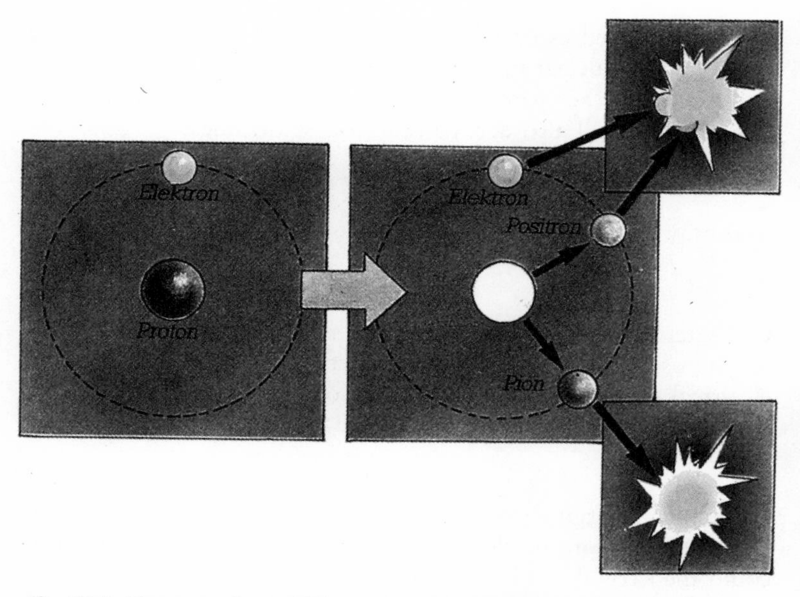

Fig. 21.2 Diagram of a possible mode of decay of a hydrogen atom. The proton decays into a positron and a neutral meson, which in turn decays into two photons. The electron and the positron also annihilate into two photons. In the end, the entire atom has been transformed into electromagnetic radiation energy.

with that lifetime, all you need do is look at a large enough number of protons. If we could observe several thousand tons of water in detail, we should be able to detect the decay of several dozen protons in one year. Experiments like this have in fact been done in various places in the world—usually deep down mine shafts, where the detectors can be shielded against cosmic radiation.

To date, not a single decay of a proton has been observed, but in the meantime research has established quite a respectable limit for the lifetime of a proton: at least 10^{31} years. In other words, Einstein, you may rest assured that if matter does in fact decay—and few serious particle physicists doubt that it does—the process occurs at such a low rate that there is no reason to be alarmed. Matter appears to decay very, very slowly.

EINSTEIN

Still, if you're right, we can easily predict the fate of our world in some distant future: all matter will be converted to radiation.

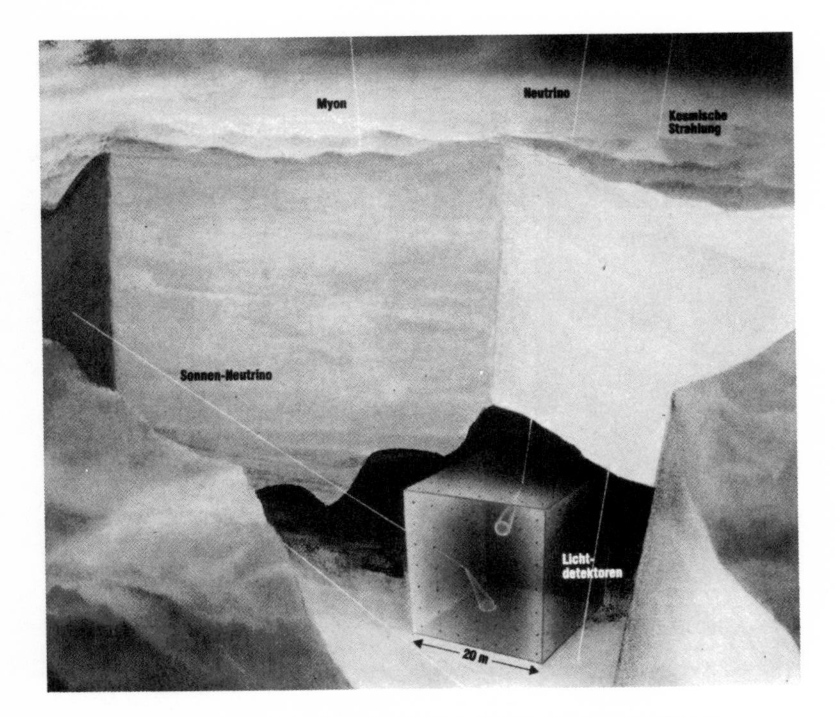

Fig. 21.3 Schematic representation of a particle detector used in the search for proton decay and for other rare processes. One such detector was built in the Morton Salt Mine near Cleveland, Ohio. Some 10,000 tons of water fill a cubic volume containing about 10^{33} protons. Every one of these protons may have a minuscule chance to decay sometime during the years of observation. During the decay process, the resulting charged particles are expected to emit blue light, known as Cerenkov radiation (named after the Russian physicist P. A. Cerenkov). This radiation will be registered on the edge of the tank by light-sensitive detectors. Although installed deep underground, detectors of this kind cannot fully escape background signals from cosmic radiation. Neutrinos and certain highly energetic muons can penetrate the shield of all the matter above the apparatus and may initiate a reaction in the sensitive volume. Some of these background reactions may be due to solar neutrinos. (Courtesy of GEO/ Joerg Kühn.)

My formula on the equivalence of energy and mass will determine the end of our world. In that distant future, our universe will be nothing but an ocean of an infinite number of photons. There will be no stars, no galaxies. There will be nothing to indicate the rich variety of the former cosmos, nor will there be anything to show that beings like us once existed who were capable of recog-

Fig. 21.4 An inside view of the particle detector in the Morton Salt Mine near Cleveland, Ohio. The volume is filled with specially purified water. Particle interactions are viewed from the outside walls of this tank by photomultiplier tubes, which operate similarly to photocells.

In 1987, this detector registered an intense neutrino burst that originated in the supernova explosion in the Large Magellanic Cloud. A similar experiment in Japan, called the Kamiokande Detector, also registered the event.

nizing important features of the dynamics governing the universe. All that remains will be space, time, and energy. Nothing will be around that bears witness to us . . .

NEWTON

But that, gentlemen, will happen only in a very distant future, if at all!

After an intermission that permitted the excitement of the past few minutes to subside, Newton spoke again.

NEWTON

When we started these discussions a few days ago, we began with space and time. Next, there was the problem of a constant speed of light, solved with great elegance by Einstein. Then we continued step by step, a natural sequence imposing itself. We have now arrived at the point at which matter dissolves into radiation—at the ultimate demise of our universe. And it was your equation, Einstein, that acted as midwife not only for the genesis of matter in the Big Bang, but also for the extermination of all matter at some future date.

When we first met in Cambridge, Haller, all we had in mind was a brief discussion of relativity theory. We have been at it for days now, and the more we discuss, the more I realize that each new thought begets another.

In my time, when I was writing the *Principia* at Cambridge, I often wondered what the natural sciences would come to. What I had in mind was a closed edifice of thought, capable of explaining all that we see around us. Today, looking back, I recognize the illusion I was indulging in. Never could I have imagined that the natural sciences—of which, after all, I'm a founding father—would develop in this manner; never could I have hoped that they would one day become as interesting and wonderful as today's physics. I'm sure I can speak for my colleague Einstein too when I offer you and your colleagues my very best wishes.

We were walking down the corridors of CERN toward the office occupied by one of my colleagues who had promised to show our visitors a couple of the experimental installations. I was going to slip away to the phone to arrange for our departure—a taxi to take Einstein to the railway station, from which he could take the train to Bern; another taxi to take Newton to the airport. So I hurried back to my office and . . .

Epilogue

" . . . and did what?" Haller and I were still sitting on the beach of the El Capitan State Park in California, watching the Pacific breakers roll in. Haller had broken off his narrative.

"So what happened next?" I asked. "Did you see Einstein and Newton again? And I really mean did you 'see' them again?"

"Of course not. I had left Newton and Einstein to go to my office. As soon as I went in, I felt bright sunlight on my face. The sunlight woke me up—still in Cambridge, still lying on the lawn. The sun was in its zenith—I must have slept for several hours and dreamed with an intensity I'd never known before. For the rest of that day I could think of nothing but Newton and Einstein. When I walked across the Trinity quadrangle again in the afternoon, I even caught myself furtively watching for someone who looked like Newton. But I was out of luck. My dream had run its course."

Sources of Quotations

Page vi
"Most books about science . . ." from *Albert Einstein: Briefe* (Zurich, 1981), p. 41.

Page xi
"Einstein explained . . ." from Chaim Weizmann, *Einstein: The Human Side* (Princeton: Princeton University Press, 1948), p. 62.

Pages 3–4
"His peculiar gift . . ." from Emilio Segrè, *From Falling Bodies to Radio Waves* (New York: W. H. Freeman, 1984), pp. 49–50.

Pages 6–7
"Hitherto, we have explained . . ." from Isaac Newton, *Principia,* trans. Andrew Motte, rev. ed. (Berkeley and Los Angeles: University of California Press, 1934), pp. 546–47.

Page 13
"Absolute, true, and mathematical time . . ." from Newton, ibid., p. 6.

Pages 42–52
"Newton's Reading: What Is Light?" originally published as an article by Harald Fritzsch, "Photonen machen die Erde hell," *PM* 12 (1984).

Page 183
"If an object gives off . . ." from Albert Einstein's paper in *Annalen der Physik* 18 (1905): 639.

Page 203
"And then without a sound . . ." from Peter Goodchild, *J. Robert Oppenheimer: Shatterer of Worlds* (Boston: Houghton Mifflin Co., 1981), p. 164.

Page 203
"A few people laughed . . ." from Goodchild, ibid., p. 162.

Page 205

"If atomic bombs are to be added . . ." from David Hawkins, *1947 Manhattan District History Project Y: The Los Alamos Project* (Los Alamos Scientific Laboratory, vol. 1), cited in Goodchild, ibid., pp. 172–73.

Suggested Reading

1. A selection of books that attempt to present the special theory of relativity in a manner accessible to the general reader

H. Bondi. *Relativity and Common Sense*. New York, 1964.
M. Born. *Einstein's Theory of Relativity*. New York, 1962.
A. Einstein. *Relativity: The Special and the General Theory*, 17th ed. New York, 1961.
S. Hawking. *A Brief History of Time*. New York, 1988.
G. Gamow. *One, Two, Three . . . Infinity*. New York, 1965.
M. Gardner. *The Relativity Explosion*. New York, 1976.
S. Goldberg. *Understanding Relativity*. Oxford, 1984.
E. Harrison. *Cosmology*. New York, 1981.
S. Lilley. *Discovering Relativity for Yourself*. New York, 1980.

2. A detailed historical evaluation of Einstein's work

A. Pais. *Subtle Is the Lord: The Science and Life of Albert Einstein*. New York, 1984.

3. Introductory works on particle physics and cosmology for the general reader

H. Fritzsch. *Quarks: The Stuff of Matter*. New York, 1989.
———. *The Creation of Matter: The Universe from Beginning to End*. New York, 1988.
S. Glashow. *The Charm of Physics*. New York, 1991.
L. Lederman. *The God Particle*. Boston, 1993.
A. Pais. *Inward Bound: Of Matter and Forces in the Physical World*. New York, 1986.
S. Weinberg. *The First Three Minutes: A Modern View of the Origin of the Universe*. New York, 1976.
———. *Dreams of a Final Theory*. New York, 1992.

Glossary

Acceleration Change in speed of a moving object, per unit of time.

Alpha particle Nucleus of the helium atom, composed of two protons and two neutrons. Alpha particles are emitted by various radioactive substances (alpha radiation). Abbreviated as α-particle.

Antimatter Matter consisting of the antiparticles to ordinary matter—antiprotons and antineutrons in the nucleus, and positrons in the surrounding shell.

Antiparticles There is an antiparticle for every particle in nature, of the same mass but of the opposite charge. The antiparticle for the (electrically negative) electron is the (electrically positive) positron, etc. Some neutral particles are identical to their antiparticles, e.g. the photon and the neutral pion.

Atom Matter as we know it consists of atoms. These in turn are composed of an electrically positive nucleus and a shell. The nucleus consists of protons and neutrons, generically known as nucleons; the shell is made up of electrically negative electrons. Atomic size is defined by the shell; the nucleus at its center is, by comparison, much smaller—by a factor of about two thousand in diameter. Yet the nucleus contains almost all the mass of the atom.

Beta decay The decay, or disintegration, of a neutron—either free or bound in a nucleus—into a proton, a neutrino, and an antineutrino. It is caused by so-called weak nuclear interaction. Abbreviated as β-decay.

CERN Acronym for Conseil Européen pour la Recherche Nucléaire. Founded in 1954 by twelve European nations as a common research laboratory for elementary particle physics, it is today the largest international research laboratory in the world.

Cosmic radiation A term applied to the radiation of particles produced in collision with other particles impinging from distant sources in the universe on the Earth's upper atmosphere. It consists mainly of protons, neutrons, light nuclei, and pions. The pions have a short lifetime and decay during their flight to Earth into photons or muons and neutrinos.

Decay Many particles, on both the nuclear and the subnuclear levels, are unstable. They eventually disintegrate into several daughter particles, which may in turn be stable or unstable. This process of disintegration is commonly called decay.

DESY Acronym for the Deutsches Elektronen Synchrotron Laboratory in Hamburg, Germany. It is the German center for particle physics research. Its most recent flagship accelerator is the storage ring HERA, containing both electrons and protons in opposite orbits which can be steered into collision at predetermined crossing points. HERA began operating in 1992.

Deuteron Particle consisting of a proton and a neutron, the nucleus of "heavy hydrogen" or deuterium.

Electrodynamics Scientific discipline dealing with electromagnetic forces and phenomena in nature.

Electromagnetic force A general term for the forces that act between electrically charged objects or particles. A special case is the attraction or repulsion observed between equally or oppositely charged objects. Electromagnetic forces are mediated by the electromagnetic field.

Electron The lightest elementary particle that has an electric charge. Electrons make up the charged shell or cloud that surrounds atomic nuclei. Their charge is defined as the unit charge; it is equal but of opposite sign to that of the proton.

Elementary particle Hundreds of elementary particles besides the constituents of the atoms are known today. Most, however, are not truly elementary but are composed of even smaller constituents called quarks. All particles so far observed can be reduced to six quarks and six leptons (the latter are related to electrons). Normal matter contains only two quarks (denoted by *u* and *d*) and electrons.

Energy A quantity defined as the ability to do work. It may appear in various forms, of which one is the energy of motion, kinetic energy. According to relativity theory, energy and mass can be transformed into each other. Energy is measured in units of joule (J) or watt-seconds (Ws). Sometimes the outdated unit erg (10^7 ergs = 1 watt-second) is used. For daily life, the unit kilowatt-hours (kWh) is appropriate. In atomic or nuclear physics, a unit called the electron volt (eV) is commonly used. It is defined as the energy acquired by an electron when it "drops" through the potential difference of one volt.

Ether A hypothetical medium, the existence of which has been postulated in various ways in order that long-range forces such as gravitation and electromagnetism might be reduced to short-range interactions. The concept of an ether at rest throughout the universe is closely akin to Newton's idea of absolute space and absolute time, both independent of the observer's reference system. Ether was also invoked to describe the spreading of electromagnetic fields and waves. Modern physics rejects the notion of ether. Instead, the forces observed are seen as a manifestation of gravitational and electromagnetic fields.

Galaxy A major grouping of stars, which may contain up to a thousand billion (10^{12}) stars held together by the forces of gravity. Elliptic, spiral, beam-shaped, or other (irregular) galaxies have been observed.

Gravitation The weakest force known in nature. It originates in the mass of an object. All masses attract each other, and the strength of this attraction depends both on the masses and on their relative distances.

Half-life The time within which one-half of a radioactive substance will decay. For uranium 238, this time amounts to some 4.5 billion years; for tritium, 12.3 years; for strontium 89, only 50.5 days. The element cesium 137 has a 30-year half-life. In particle physics, particle decay is usually described as lifetime τ, or mean life. Half-life $t_{1/2} = 0.693\tau$.

Inertial system A physical reference system in which the motion of a free object is described by a straight line.

Lifetime *See* Half-life.

Light year A unit of length used in astronomy; it corresponds to 9.46×10^{12} km, the distance traveled by light in one year.

Mass A basic quantity in physics, mass is responsible for the inertia of a physical object against any change of its state of motion, and for its weight in the gravitational field of other bodies. According to relativity theory, the mass of an object depends on its state of motion. The equivalence of mass and energy was first recognized by Einstein.

Momentum A term for the quantity of motion of an object or body; it is the product of mass and speed.

Muon An elementary particle related to the electron, but with a mass about 200 times larger. The muon is unstable and after a brief time decays into an electron, a neutrino, and an antineutrino.

Neutrinos Neutral partners of the electron; to this day, we know of the existence of three neutrino species—electron-neutrinos, muon-neutrinos, and tau-neutrinos. The symbols are ν_e, ν_μ, and ν_τ.

Neutron An electrically neutral particle which, together with the proton, is one of the building blocks of atomic nuclei. An unbound neutron is unstable; it decays into a proton, an electron, and a neutrino.

Nuclear fission The splitting of a heavy atomic nucleus into two or more lighter ones. Nuclear fission can occur spontaneously, like radioactive decay; it can also be induced by bombardment by particles such as neutrons. The fission of heavy nuclei can release energy.

Nuclear force The force responsible for binding nucleons in an atomic nucleus. It is due to strong interaction among quarks. Although it is the strongest force occurring in nature, it only acts over distances too small to allow direct observation in macroscopic experiment.

Nuclear fusion During fusion, two nuclei, usually of small size, combine to form a heavier one. In the process, a large amount of energy is released, a part of the total mass being transformed into energy due to the equivalence of mass and energy. This process fuels energy production in the stars.

Nuclear interaction There are two types of nuclear interaction forces. The strong nuclear force is responsible for nuclear binding. The weak nuclear force, of greatly reduced strength, is responsible for the disintegration, or decay, of nuclei or elementary particles with longer lifetimes. These lifetimes may range from tiny fractions of a second to many years.

Nucleon The constituents of an atom's nucleus—protons and neutrons.

Particle accelerator Machine that accelerates electrically charged particles, mostly electrons or protons, to very high energies. The process utilizes electromagnetic fields. Circular accelerators accelerate particles in ring-shaped vacuum tubes; linear accelerators, in a straight vacuum tube. The largest circular accelerator to date is the LEP machine at CERN, with a circumference of 23 km. It has been operating since 1989.

Photon The quantum, or "particle," of light, abbreviated as γ. It has no rest mass and therefore always moves at the speed of light.

Pion The pion, also called π meson, is an unstable elementary particle. It is involved in the strong nuclear interaction. It has three charge states, π^+, π^-, π^0. Pions are now known to be the lightest particles that consist of a quark and an antiquark.

Plasma When matter is heated to very high temperatures, atomic structure is destroyed by means of frequent collisions among the atoms. This results in a mixture of free nuclei and electrons, known as plasma. This is the state of matter within stars such as the sun.

Proton The positively charged elementary particle that is also the nucleus of the hydrogen atom. All other nuclei consist of protons plus neutrons.

Quarks The constituents of nucleons (i.e., protons and neutrons). The existence of six different types of quarks has been directly or indirectly established. Their symbols are *u, d, c, s, b,* and *t.* Particles containing *a, d, c, s,* and *b* quarks have actually been observed, while particles containing *t* quarks have not yet been established. It is thought impossible for quarks to be found as isolated particles, since the forces binding quarks increase steadily as the distance between them is increased.

Radioactivity Atomic nuclei that emit other particles spontaneously are known as radioactive. Three kinds are recognized, depending on the type of particle emitted: alpha (α) radiation denotes the emission of helium nuclei; beta (β) radiation consists of electrons; whereas gamma (γ) radiation consists of photons. Radioactivity can be dangerous because it can destroy vital molecules in living substances, such as the composition of genes.

Space-time The name used for the unified concept of space and time that is a vital part of relativity theory. In space-time, the coordinate system has four dimensions—three spatial and one temporal.

Speed *See* Velocity.

Storage ring A ring-shaped machine that can store circulating elementary particles that have been accelerated to high speeds.

Supernova An exploding star, which ejects the majority of its matter into interstellar space. Such explosions liberate as much energy as is radiated by the sun in several billion years. The last supernova observed in the Milky Way was described by Johannes Kepler in 1604. In 1987, a supernova was observed in the Large Magellanic Cloud, one of the small galaxies in the vicinity of the Milky Way.

Velocity The rate of motion of an object, usually measured in units of centimeters per second (cm/s), meters per second (m/s), or kilometers per second (km/s). While we use the terms velocity and speed (as in speed of light) interchangeably in this text, the proper physics definition of velocity includes the direction of motion, while that of speed does not.

Weak nuclear force A very weak force that acts between elementary particles such as electrons, neutrinos, and quarks. It is responsible for β radioactivity. This force is mediated by other elementary particles, "weak bosons," which are about one hundred times as heavy as protons. For this reason, weak nuclear force acts only over very short distances.

World line In space-time, a moving object defines a line—the sequence of events through which the object passes. One such line is known as a world line; it contains all the information about the motion of an object in the past, the present, and the future.

Index

Italicized page numbers refer to illustrations.

absolute space, 13–15, 18, 21–22, 30–35
absolute time, 13, 18, 30–32, 73
acceleration
 absolute nature of, 30
 antimatter and, 222
 inertial forces and, 21
 mass and, 168, 173–75
 relativity theory and, 85
 speed of light and, 85
 time dilation and, 141–43
 See also particle acceleration
Adams, John Couch, 8
aging, twin paradox and, 139–44
Alamogordo, 202
alpha particles, 190–92, *192*
Anderson, Carl, 226–27, *227*, 229, 236
Andromeda galaxy, 34, *137*, 143
Annalen der Physik, 55, 182–83, 206
Anne, queen of England, 9
annihilation, 232–34, 236, 238
 of antiprotons, *241*, 242
 electron-positron, 229–31, *231*, 239–42, *240*, *241*, 246, 249, *250*
 quark-antiquark, 249
antideuterons, 228
antielements, 228
antihelium, 228
antihydrogen, 228–29, *229*, 233
anti-iron, 228, 233
antimatter, 222–34
 matter and, 233–34
 in our universe, 235–37
 particles in, 246–49
antineutrons, 228
antinuclei, 228
antiparticles, 222, 225–28, 236
antiprotons, 222, 227–28, 236, 239, 246
antiquarks, 246–49, *249*
antistars, 235–36
Aristotle, 11–12
ASDEX combustion chamber, *214*
astronomical phenomena, Newton and, 6–8
atomic bombs
 at Hiroshima, xiv
 Manhattan Project and, 202–5
 nuclear fission and, 200–201
atomic clocks, 101–2, *102*, 135
atomic nuclei
 atomic theory and, 39
 barium, 197–98, *198*
 electrical forces and, 39–40, 43
 fission of, 196–201
 gold, 197, 242–43, *243*
 heavy, 197
 heavy hydrogen, 184–86, *186*, 187
 helium, 190–91, *191*, *192*, 194–95, 210
 hydrogen, 250
 iron, 197
 krypton, 197–98, *198*
 muon creation and, 124–25
 numbers of, 52
 stability of, 248
 sulfur, 242–43, *243*
 unstable, 217–18
 uranium, 197–200, *198*
atomic processes, space-time and, xv
atomic structure, *41*. *See also* atomic nuclei
atomic theory, 39–40
 classical mechanics and, 8
 density of matter and, 10
 electrical charges and, 43
 measurement of time and, 101–2
 photons and, 48
 probability and, 127–28
 relativity theory and, 223–25, 229
 strong interaction and, 40
atomic weapons, use of term, 200. *See also* nuclear weapons
atoms
 atomic theory and, 39–40

atoms (*continued*)
deuterium, *40*
helium, 190–91, *191*
high temperature and, 194
incident light and, 44
iron, 197
measurement of time and, 101–2
Newtonian view of, 10
size of, 149–50
space contraction and, 149–50
stability of, 149–50
visualization of, *41*
See also atomic nuclei
Augustine, Saint, quoted, 12, 112

background radiation, 51–52, 251
barium, 197–98, *198*
Barrow, Isaac, 3–4
Bern, Einstein's home in, 58–59
Big Bang
antimatter and, 237–38
background radiation and, 52
creation of matter and, 248, 253
relativity theory and, xiii
special theory of relativity and, xvi
strong interaction and, xiv–xv
Bohr, Niels, 128
bubble chamber, *241*

cadmium, 212
calculus
differential, 3
variational, 5–6
calibration, of units of length, 103
California Institute of Science and Technology (Caltech), 226
Cambridge University, Newton at, 1–3, 5
Cerenkov radiation, 251
CERN. *See* European Laboratory for Particle Physics
cesium, 101, *102*
chain reactions, nuclear, 199–200, 211
Charles II, king of England, 9
Chernobyl, accident at nuclear power station, 215–16
circle, in space-time, *80*
classical mechanics, xii
acceleration and, 85
Einstein and, 9
electromagnetic processes and, 8
energy and, xv, 178
light and, 36–38, 42–43, 53–54, 64, 70–71
mass and, xv, 95, 165, 167–68
Newton and, 3, 5–9, 29–35
space and, 12–18, 30–35, 90
speed and, 64
speed of light and, 85, 104
time and, 12–13, 15–18, 89–90
clocks
atomic, 101–2, *102*, 135
light, 107–11, *109*
synchronization of, 105–7
time dilation and, 107–11, 121, 134–36
cloud chambers, 226
coal, toxic waste from, 216–19
collisions. *See* particle collision
Coma cluster of galaxies, *33*
Compton, Arthur H., 48–50
Compton effect, 50
conservation
of energy, 190
future energy production and, 219–21
conservation laws, xv
constant, of relativity theory, 154–63
constellations, *33*, *34*
coordinates
motion and, 74
relativity theory and, 159–60, 162
space, 73–74
space-time, 159–60, 162–63
time as, 73–79
coordinate systems
Cartesian, *14*
description of space and, 72–79
distance within, 154–55, *155*
motion and, 16–21, 35, 77–78
space and, 13–15, 29–30
time dilation and, 108–14
See also inertial systems; moving systems; rotating systems; space-time coordinate systems
cosmic background radiation. *See* background radiation
cosmic rays, 124–25, 129–30, 226
critical mass, 200, 212
curvature, of particle tracks, *172*, 174, 226
curves
in space-time, 78, *80*
as world lines, 77, 141

decay. *See* particle decay
deceleration, time dilation and, 141–43
density, mass and, 10
DESY. *See* German Electron Synchrotron

deuterium, *40*, 184–88, 193–94, 210–11
deuterons
  in alpha particle, 190–92, *192*
  in heavy hydrogen, 187–88
  mass of, 184–86, *186*
  nuclear fusion and, 190–94, 210
Dialogue on the Two Chief World Systems (Galileo), xvii
differential calculus, 3
dimensions
  space-time, 75–79, 81–83, *82*, 92, 163
  spatial, 13–15, 72–73, 163
  time, 73, 81, 163
Dirac, Paul A., 223–26
  theory of electrons, 225, 228–31
Dirac function. *See* equations, Dirac
directions
  space contraction and, 149
  spatial, 13–15, 17–18
  speed of light and, 91
  time and, 17–18
distance
  within coördinate systems, 154–55, *155*
  gravitation and, 96
  measurement of, 100–101, 103, 149–51
"Does the Inertia of an Object Depend on Its Energy?" (Einstein), 55

Earth
  inertial systems on, 20–21
  nuclear fusion on, 196
Einstein, Albert, 42, *62*, *67*
  achievements, 3–4
  biographical sketch, 54–55
  "Does the Inertia of an Object Depend on Its Energy?" 55
  "Electrodynamics of Moving Objects," 55
  energy-mass transformation and, xii–xiv
  gravitation and, 3
  "Heuristic View of the Generation and Transformation of Light, A," 55
  house, 58–59, *60*, *61*
  light and, 38–39, 42–53
  mechanics and, 9
  quoted, 4–5, 46, 183, 204, 217
  See also equation, energy-mass; relativity theory
Einstein, Hermann, 54
  house of, *54*
Einstein Society of Bern, 58
electric charge
  annihilation and, 232
  antimatter and, 225, 227–28
  atomic theory and, 43
  light waves and, 94–95
  of mesons, 246, 247
  muon decay and, 124–26, *126*
  nuclear fission and, 197
  nuclear fusion and, 192–93
  reciprocal attraction and, 93
  transfer of energy and, 50, 52
electric fields
  disappearance of, 96–97, 99
  electromagnetic waves and, 92–94, *93*
  light waves and, 94–95
electric forces
  atomic theory and, 39–40, 43
  photoelectric effect and, 44–47
electric repulsion
  nuclear fission and, 197
  nuclear fusion and, 192–93
"Electrodynamics of Moving Objects" (Einstein), 55
electromagnetic fields, 92–98
electromagnetic processes
  electron-positron pairs and, 232
  mechanics and, 8
  particles and, xv
electromagnetic radiation
  annihilation and, 229–30
  energy-mass conversion and, 183, 185, 188–89, 192
  nuclear fission and, 199
  from sun, 181
electromagnetic signals, speed of, 107–8. *See also* speed of light
electromagnetic waves
  light and, 43–44, 64–69
  space-time and, 93
  X-ray, 43
electron-positron pairs, 231–32, *232*
electrons
  acceleration of, 222
  annihilation of, 229–31, *231*, 238–39
  atomic theory and, 39–40, *40*, 43
  Dirac equation and, 225–26
  in helium atom, 190
  high temperature and, 194
  magnetic fields and, 225
  mass of, 49–50
  muon decay and, 125
  number of, 52

electrons (*continued*)
particle collision and, 48–50, 229–31, 238–40
photoelectric effect and, 44–47, *44*, *45*
space contraction and, 151
speed of, 49
time dilation measurement and, 122
electron volts, 173
measurement of mass in, 185
energy
acceleration and, 85, 171
conservation of, xv, 190, 219–21
creation of deuterium and, 187–88
frozen, 180–81
kinetic, 175, 178–79, 193, 199, 210
mass and, xii–xiv, 175–90
measurement of, 183, 187
mechanics and, xv
meson decay and, 248
motion and, 11, 177
nuclear fission and, 196–97
nuclear fusion and, 191–94, *192*
particle acceleration and, 174–75
particle annihilation and, 229–31, 242
particle collision and, 239–43
particle creation and, 231, 239
particle decay and, 250
of particles, 173
photoelectric effect and, 44–47
in photons, 46–47, 50, 52, 175
relativity theory and, 183
rest mass and, 179–80
speed and, 177–80
speed of light and, 180
from sun, xiv–xv, 181, 193–95, 220
See also equation, energy-mass
energy production
annihilation and, 233–34
conservation and, 219–21
future, 219–21
from nuclear fission, 211–19
nuclear fusion and, 209–11
equation, energy-mass, 177–89
impact of, xi–xii
heavy particle production and, 239–42
matter decay and, 251–53
nuclear fission and, 196–99
nuclear fusion and, 194–96
nuclear weapons and, 204–7
particle annihilation and, 230–31
equations
Dirac, 225, 228–31
distance within coordinate systems, 154
gamma factor, 138. *See also* gamma factor (Lorentz factor)
kinetic energy, 175–76, 178–79
mass-energy transformation and, xii–xiv
proper distance, 159
relativistic mass, 171, 173
relativistic mass-energy, 178–79
time dilation, 113–15
ether
light waves and, 38, 43
as medium for electromagnetic waves, 64–69
Michelson-Morley experiment and, 91–92
European Laboratory for Particle Physics (CERN)
antiprotons at, 236
experiments at, 130–33, *131*, 151–53
particle accelerators at, *57*, 85–86, 173–75, 222, 224, *225*, 228
events, 76
lightlike, 160–61
spacelike, 160–61
space-time coordinates for, 79
timelike, 160–62
experiments
Michelson-Morley, 66–69, 88–89, 91–92
space contraction, 151–53
time dilation, 120–33
experiments, thought
gravitational waves and, 96–99
mass and, 168–71
relativity theory constant and, 155
speed of light and, 87–89
time dilation and, 134–35, 139–43, 145–48

Faraday, Michael, 53
Federal Insititute of Technology (ETH), 54–55
Fermi National Accelerator Laboratory (Fermilab), 236
flow of time
matter and, 136
motion and, 112–13
observer and, 149
time dilation and, 118
fossil fuels, waste from, 216–17
Frisch, Otto Robert, quoted, 203

Index

galaxies, 237
  motion of, 33–35
  See also Andromeda galaxy; Coma cluster of galaxies; Large Magellanic Cloud
Galileo
  Dialogue on the Two Chief World Systems, xvii
  motion and, 11–12
Galle, Johann Gottfried, 8
gamma factor (Lorentz factor)
  calculation of, 114–15
  divided by gamma factor, 154
  mass and, 171
  mass-energy equation and, 178
  muon decay and, 132–33, 146
  particle decay and, *156,* 158
  particle lifetime and, 158
  space contraction and, 148, 150, 154
  time dilation and, 114–18, 134–35, 137–38
gamma quanta, 47, 217, *231. See also* photons
gamma radiation, 47, 236
general relativity theory, xii, 55. *See also* relativity theory
geometry, space-time and, *75,* 92
German Electron Synchrotron (DESY), 239–40
German Federal Institute of Technical Physics, 102
God, 22, 30–31
gold, 242–43, *243*
gravitation
  general relativity theory and, 55
  mass attraction and, 95–96
  Newton and, 3, 6–8
  speed of light and, 99
gravitational field, 96–99
gravitational waves, 96–99

Habicht, Conrad, 62
half-life. *See* lifetime
Haller, Friedrich, 166–67
Halley, Edmund, 5
heating, nuclear fusion and, 193–94
helium, 190–91, *191*
  nuclear fission and, 196
  nuclear fusion and, 194–95, 210
Hermann von Helmholtz Institute, 67
Hertz, Heinrich, 43–44, 53
"Heuristic View of the Generation and Transformation of Light, A" (Einstein), 55
Hiroshima, xi, xiv, 203, 205
homogeneity
  of space, 14–17
  of time, 16–17
Huygens, Christian, 43
hydrogen, 194–95
  decay of, 249–50, *250*
  heavy, 184–88, *186,* 194
  helium production and, 194
  nuclear fission and, 214
  proton decay and, 249
hydrogen bomb, 208–9

inertial forces
  acceleration and, 21
  space and, 30
inertial systems, 29–30
  absolute space and, 22
  on Earth, 20–21
  motion and, 19–22, 29–33
  time dilation and, 140–41
inertia principle, 12
  motion and, 16, 18–20
  speed and, 76
integral calculus, 3
interference phenomena, 69
iron, 197
isotropy, of space, 15

JET. *See* Joint European Torus Lab
Joint European Torus Lab (JET), 211–13, *212, 213*

Kaiser Wilhelm Society, 55
Kamiokande Detector, 252
krypton, 197–98, *198*

Large Electron Positron (LEP), *57,* 222, 224, *225*
Large Magellanic Cloud, 34, 96, *98,* 99, 252
laser beams, 106, 211
length, measurement of, 100–101. *See also* distance
LEP. *See* Large Electron Positron
Le Verrier, Urbain J. J., 7–8
lifetime
  of mesons, 246–48
  of muon, 125–28, *127*
  particle, 155–58, *156, 157*
  of protons, 248–50
light
  Einstein and, 47–49, 53
  as electromagnetic field, 92, 94
  incident, 44–46
  laser, 106, 211

light (*continued*)
  Newton's theories of, 36–38, 42–43, 53–54, 64
  as particle. *See* light particles; photons
  particle-wave dualism and, 40, 42–52
  relativity theory and, 71
  space-time coordinate systems and, 79–81
  speed of. *See* speed of light
  as wave. *See* light waves
  wavelengths, 43, 46–47, 251
  wave theory and, 38–39
  See also photons; speed of light
light clocks, 107–11, *109*
light cones, 81–83, *82*
light particles, 36–38, 42–43, 46–47. *See also* photons
light second, use of term, 80
light signals
  light clocks and, 107–11
  lightlike events and, 160–61
  measurement of distance and, 151
  measurement of time and, 105–6
  space-time and, 87–89, *88*
  world lines for, *81*
light waves, 43–45, 64–66, 94
light year, use of term, 80
Lorentz factor. *See* gamma factor
Los Alamos, 201, 202
Lucas, Henry, 3

Magic Mountain, The (Mann), 12
magnetic fields, 92–94
  electrons and, 225
  particle acceleration and, 174
  particle tracks and, *172,* 174, 226
Manhattan Project, 202–5
Mann, Thomas, *The Magic Mountain,* 12
mass
  acceleration and, 168, 173–75
  annihilation and, 229–31
  of antimatter, 227–28
  of antiprotons, 227
  attraction of, 95
  conservation of, xv, 190
  converted into energy, 198–99
  critical, 200
  density and, 10
  of deuterons, 184–86, *186*
  of electrons, 49–50
  of elementary particles, 49–50
  energy and, 175–90, 198–99
  of helium, 190–91
  measurement of, 165, 185
  mechanics and, xv
  moving, 171–74, *172,* 177–78
  of muons, 123
  of neutrinos, 86–87
  of neutrons, 184–86
  Newton's concept of, 165, 167–68
  nuclear fission and, 198
  nuclear fusion and, 198
  particle collision and, 231
  of particles, 49–50
  photons and, 175
  of pi-mesons, 246
  of positron, 227
  of protons, 178, 184–86
  of radium, 217
  relativity theory and, 165, 167–76
  rest, 171, 173, 175, 178–80
  speed and, 168–71
  speed of light and, 171–73
  of sun, 195
  transformed into energy, xii–xiv
  volume and, 10
  See also equation, energy-mass
mass deficit, 186–87
massive objects, motion of, 29–30
massive point, motion of, 15–16, *17*
material objects, speed of, 84–85
matter
  antimatter and, 228, 233–34, 235–38
  creation of, 252
  decay of, 245–53, 250–53
  generated by Big Bang, 248, 253
  mesons and, 246–47
  Newtonian view of, 10
  in our universe, 236–37
  relativity theory and, 163
  stability of, 149–50
  time and, 136
  See also particles, elementary; *names of particles*
Maxwell, James Clerk, *Treatise on Electricity,* 101
measurement
  of distance, 149–51
  of energy, 183, 187
  of length, 100–101
  of mass, 165, 185
  of particle decay, 155
  of particle lifetime, 155
  of particle trajectories, 222
  of space, 18, 160
  of space contraction, 149–51
  of speed of light, 66–69. *See also* experiments, Michelson-Morley

of time, 12–13, 18, 31, 100–102, 105–6
of time dilation, 113–15, 121–33, 134–35
mechanics. *See* classical mechanics; quantum mechanics
mega electron volts (MeV), as measure of mass, 185
meltdown, 214
mesons, 240–41, 245–49, *249*
antiquarks in, 246–47, *249*
decay of, 246–48
mass of, 246
neutral, 246, 248, 250
particle collision and, 243
pi, 242, 245–46
quarks in, 246–47, *249*
upsilon, 239–40, *240*
meter, standard, and speed of light, 100–101
Michelson, Albert Abraham, 66–69, *67*
Milky Way, *34*
antimatter in, 235–36
motion of, 34
Millikan, Robert A., *67*
Minkowski, Hermann, 160
momentum, mass and, 169–71, *169*
Morley, Edward Williams, 66–69
Morton Salt Mine (Cleveland), *251*, *252*
motion
absolute, 18–19
Aristotle and, 11–12
coordinate systems and, 18–21
cosmic, 33–35
energy and, 177
flow of time and, 112–13
galactic, 33–35
Galileo and, 11–12
inertial systems and, 19–22, 29–30
inertia principle and, 11–12, 18–20
measurement of time and, 12–13
Newtonian view of, 5–6, 10–13
periodic, 12
of photons, 49
planetary, 6–8, 20
relative, 18–19
relative space and, 13
relativity theory and, 103–4, 155, 161–62
of rigid bodies, 10, 12
space contraction and, 148
space-time and, 74–79
speed of light and, 83–85
time and, 15–17
time dilation and, 108–13, 134–36
world lines and, 76–79
moving systems
coordinates and, 18–20
motion of, 16
relativity theory and, 161–62
space-time continuum and, 74–79
speed of light and, 83–85, 87–89
time dilation and, 108–11, 113, 130, 134–36, 139–42
world lines for, 76–79
muons, 245
decay of, 123–29, *124*, *126*, *132*, 145–48
generation of, 124–45, 245
proton decay research and, 251
speed of, 129, 145
time dilation and, 123–33, 145–48

Nagasaki, 203
neutrinos
mass of, 86
muon decay and, *124*, 125
proton decay research and, 251–52
neutrons
antiparticles of, 228
atomic theory and, *40*
converted into helium, 194–95
in helium nucleus, 190
mass of, 184–86
nuclear fission and, 196–200, *198*, 212
nuclear fusion and, 195, 210
in uranium nucleus, 197–98, *198*
Newton, Sir Isaac
absolute space and, 13–15, 18, 21–22, 30–35
absolute time and, 13, 18, 30–32, 73
achievements, 3–4
on British currency, *8*
calculus and, 3, 5–6
at Cambridge, 1–3, *2*, 5
fame, 9
gravitation and, 6–8, 95–96
light and, 36–38, 42–43, 53–54
mass and, 165, 167–68
on matter, 10
mechanics, xii, xv, 3, 12–13, 29–35, 36–38, 42–43, 64, 89–90
momentum and, 10
motto, *Hypotheses non fingo,* 7–8, 22, 163
Opticks, 4–5

Newton, Sir Isaac (*continued*)
Philosophiae Naturalis Principia Mathematica, 5–10, 12, 16–17, 18, 30, 37, 90, 95
principle of inertia, 12, 16, 18–20
principle of relativity, 104–5
quoted, 5, 6–7, 18, 31
religion and, 22, 30–31
research methods, 5
space and, 12, 18, 31, 71, 90, 154–55
time and, 12–13, 15–18, 71, 89–90
Nobel Prize, 55
nuclear energy
from fission, 211–19
from fusion, 209–11, 220
strong interaction and, xiv–xv
nuclear explosion, 199–200. *See also* nuclear weapons
nuclear fission, 196–201
efficiency of, 198–99
energy production and, 211–19
Manhattan Project and, 202
nuclear fusion and, 208–9
nuclear weapons and, 208–9
radioactive waste from, 216–19
See also energy
nuclear forces
heavy hydrogen and, 184, 187
nuclear fusion and, 192–93, 196
See also atomic nuclei
nuclear fusion, 191–94
on Earth, 196
energy and, 191–94, *192*
energy production and, 220
nuclear weapons and, 208–9
in sun, 193–95, 209–10, 220
See also energy
nuclear reactions
chain, 199–200, 211
energy from sun and, 190–96
gamma radiation and, 47
mass-energy conversion and, 190
radioactivity and, 217
nuclear reactors
fusion, 220–21
natural, 214–15
nuclear fission and, 212–15, *213*, *214*
nuclear waste and, 218–19
safety of, 214–16
nuclear stability. *See* atomic nuclei
nuclear waste
decay of, 215, 217
disposal of, 216–19
nuclear fusion and, 220–21
nuclear weapons
Manhattan Project and, 202–5
nuclear fission and, 200–201
nuclear fusion and, 208–9
nucleons
atomic theory and, *40*
nuclear fusion and, 210
strong interaction and, 40
nucleus. *See* atomic nuclei

object at rest
coordinate systems and, 16
energy in, 178
world line for, 76, *77*
object in motion
coordinate systems for, 16, *17*
mass of, 171–73
observer
absolute time and, 18
flow of time and, 112–13, 149
in motion, 88, 108, 112–13
moving systems and, 18–20
muon decay and, 147–48
relativity theory and, 103–4, 155, 159, 161, 168–71
at rest, 87–88, 108, 113, 139–41, 147–48, 152, 168–70
space contraction and, 148–49, *152*
speed of light and, 87–89
structure of space and, 150
time dilation and, 105, 108–13, 139–41, 146–48, 169–70
world lines for, 79
oil, 216–17, 219
Oklo (Gabon), 214–15
Olympia Academy, 59, 62
Oppenheimer, J. Robert, 202–3, 205–6
quoted, 203, 205
Opticks (Newton), 4–5

paradox
relativity theory and, 83–84
twin, 139–43
particle acceleration
antimatter and, 228
space contraction and, 151–53
particle accelerators, 85–86
antiprotons and, *228*
at CERN, *57*, 222, 224, *225*, 228
mass and, 168, 173–75

space contraction and, 151–53, *152*
at Stanford, 49, *50*
particle collision, 48–50
annihilation and, 229–30, 238–39, 241–42
antimatter and, 222, 224, *225*, 238–39
heavy particles and, 241–42
mesons and, 245–46
muons and, 245
quarks and, 247
particle decay, 122–29, *124*, *126*, *132*
heavy particles and, 239–40
measurement of, 155
mesons and, 246–48
protons and, 248–52
time dilation and, 122–33, 145–48, 155–57, *156*, *157*
particle detectors, 222, *223*, *251*, *252*
particle physics, measurement of mass in, 185
particles, elementary
atomic theory and, 39–40, 43
creation of, 231–32
decay of, 122–29, 155, 239–40, 246–52
electromagnetic processes and, xv
energy of, 173
heavy, 239–42
hybrid, 227
lifetime of. *See* lifetime
light, 36–38, 42–43, 46–47. *See also* photons
mass of, 49–50, 168, 185
muon creation and, 124–25
neutral, 227
Newtonian concept of, 10
space contraction and, 151–52
speed of light and, 86–87
strong interaction and, xiv–xv
time dilation measurement and, 122–33
See also names of particles
particle tracks, *172*, 174–75
annihilation and, 229–30, 241
curvature of, *172*, 174, 226
particle trajectories, 222
particle-wave dualism, 40, 42–52
Penzias, Arno, *51*, 52
Philosophiae Naturalis Principia Mathematica (Newton), 5–10, *6*
coordinate systems and, 30
inertia principle and, 12
mass attraction and, 95
space and time in, 16–18, 90
speed of light and, 37
Philosophical Transactions of the Royal Society, 4
photoelectric effect, 44–47, *44*, *45*
photons
annihilation and, 229–31, *231*, 238
background radiation and, 51–52
colliding with electrons, 48–50
cosmic density of, 52
decay of matter and, 251
electricity and, 50, 52
energy-mass conversion and, 187
energy of, 46–47, 50, 52, 175
as hybrid particle, 227
mass and, 50, 175
meson decay and, 248–51
nuclear fusion and, 192, 211
particle creation and, 231–32
photoelectric effect and, 44, *45*
of radio waves, 47
speed of, 49
X-ray, 47, 48–49
See also light particles
physics, Newton's. *See* classical mechanics
pi-mesons (pions), 242, 245–46. *See also* mesons
Planck, Max, xii, 47–48
planetary motion, 6–8, 20
plasma, 194
nuclear fusion and, 211, 214
plutonium, 199
bombs, 208
point mass, coordinate systems and, *17*
politics
nuclear power and, 216
nuclear weapons and, 204–5
Pope, Alexander, quoted, 1
positronium, 232–33, *233*, 246–47
positrons, 126
acceleration of, 222
annihilation and, 229–31, *231*, 238–39
discovery of, 226–27
mass of, 227
in mesons, 246
particle collision and, 229–31, 238–40
proton decay and, 248–49
power stations, nuclear, 215
Princeton University, Institute for Advanced Study, 55

Principia (Newton). See *Philosophiae Naturalis Principia Mathematica*
probability
atomic theory and, 127–28
of muon decay, 125–28, *127*, *132*
quantum theory and, 127–29
protons
acceleration of, 222
antiparticles of, 227–28
atomic theory and, *40*
decay of, 248–52
in heavy nuclei, 197
in helium nucleus, 190, *191*
lifetime of, 248–50
mass of, 178, 184–86
muon creation and, 124–25
neutron formation and, 195
nuclear fission and, 196–98
nuclear fusion and, 194–95
particle collision and, 239
quarks in, 246
relativistic mass increase and, 173–75
space contraction and, 151, *152*
speed of, 85–86
in uranium nuclei, 197–98

quantum mechanics, 223–25
quantum theory
Einstein and, 4
probability and, 127–29
quarks, 246–49, *249*
quartz crystals, 101

radiation
background, 51–52, 251
decay of matter and, 250–51, 253
mass-energy conversion and, 183
meson decay and, 248
radioactivity
natural, 217–19
nuclear fusion and, 220–21
of nuclear waste, 216–21
radio waves
electric fields and, 93–94
photons of, 47
wavelengths of, 43
radium, 183, 217
ratio, of speed of spacecraft and speed of light, 138
reference systems
relativity theory and, 160–63
rotating, 32
speed and, 21
speed of light and, 83–84, 87–89
time dilation and, 155
relative space, Newton's concept of, 13–15, 31
relativistic mass increase, 171–75, *172*
relativity
of space, 149
of speed, 21, 30, 37, 64
of time, 149
relativity principle, Newton's, 104–5
relativity theory
acceleration and, 85
in *Annalen der Physik,* 55, 182–83, 206–7
atomic physics and, 223–25, 229
energy-mass equation and, xii, xv, 177–89
experiments on, 120–33
general, xii, 55
and gravitation. *See* gravitation
importance of, xv–xvi
light and, 71
mass and, 165, 167–76
mass-energy unity and, xv
matter and, 163
motion and, 103–4
muon decay and, 130
Newton's concept of, 13
nuclear weapons and, 206–7
observer state of motion and, 155, 168–70
paradox and, 83–84, 139–43
particle acceleration and, 151–53
proof of, 183–84
quantum mechanics and, 223–25
simultaneity and, 87–90
space contraction and, 148
space-time and, xv
spatial coordinates and, 155, 159–60, 162
special, xvi–xvii, xxvi, 55
speed of light and, 84–86, 89, 104–5, 115–16
time and, 105–18
time dilation and, 107–18, 120–33
twin paradox and, 139–43
See also equation, energy-mass
religion, 22, 30–31
reversibility, of microscopic processes, 231
rigid bodies, motion of, 5–6, 10, 12
Roosevelt, Franklin D., 200–201
rotating systems
distance within, 160
inertial systems and, 32

motion of massive objects and, 30–32
spatial, 15
rotation, of space and time, 160
Royal Society, 4, 9
Russia
nuclear accidents in, 215–16
nuclear weapons and, 209

Segrè, Emilio, quoted, 3–4
shape, space contraction and, 150, 152
shock waves
electric fields and, 97
electromagnetic, 98
gravitational fields and, 99
simultaneity, relativity theory and, 87–90
Solovine, Maurice, 62
space
absolute, 13–15, 18, 21–22, 30–35
contraction of, 148–52
dimensions of, 72–73, 163
direction in, 17–18
homogeneity of, 14–17
intuitive concept of, xiii
isotropy of, 15
light and, 71
measurement of, 18, 160
Newton's concept of, 12–18, 30–35, 71, 90, 154–55
relative, 149
relativity theory and, xiii, 158–59, 162
special theory of relativity and, xii
structure of, 150
space, squared, relativity theory constant and, 158–59, 162
space coordinates. *See* coordinates
space exploration, time dilation and, 136–39
space-time
constant factor in, 154–63
as continuum, 74
coordinates for, 159–60, 162–63
dimensions of, 92
distance in, 159–63
electromagnetic fields and, 93–95
light cones and, 81–82
light signals and, 87–89
light world lines in, 79–81
quasi-rotation of, 160
relativity theory and, xv, 103
speed of light and, 86–87, 89–90, 118
structure of, 163
three-plus-one dimensional, 82–83
two-dimensional model of, 75–79, *75*
two-plus-one dimensional, 81, *82*
use of term, 159–60
space-time coordinate systems
light and, 79–81
motion and, 75–79
particle decay and, 156–58, *157*
space-time diagrams
particle decay and, 156–58, *156*, *157*
space-time constant and, 160–61, *161*
twin paradox, 141–42, *142*
special theory of relativity
birth of, 55
consequences of, xvi–xvii
energy-mass transformation and, xxvi
speed
classical mechanics and, 37, 64
of electrons, 49
energy and, 175, 177–80
gamma factor and, 113–18
inertia principle and, 76
of light. *See* speed of light
mass and, 168–71, 173–74
of material objects, 84–85
momentum and, 10
of muons, 129, 145
particle mass and, 168
of particles, 86–87
of photons, 49
of protons, 85–86
relativity of, 21, 30, 37, 64
time dilation and, 141–42
speed of light, 36–37
acceleration of mass and, 168
as constant, 37, 49, 64–71, 83–84, 86–89, 104–5
direction and, 91
electromagnetic phenomena and, 95
energy and, 180
gamma factor and, 116–18
gravitation and, 99
lightlike events and, 161
mass and, 171–73
measurement of, 66–69
mechanics and, 104
motion and, 83–85
particle collision and, 239
photons and, 175
protons and, 175

speed of light (*continued*)
reference systems and, 83–84, 87–89
relativistic mass increase and, 171, 173
relativity theory and, 84–86, 89, 104–5, 115–16
space contraction and, 149
space-time and, 86–87, 89–90, 118
time dilation and, 111–12, 115–16, 121–22, 137, 149
as unit of measure, 100–103
spiral, as world line, 77, *78*
Stanford Linear Accelerator (SLAC), 49, *50*
storage rings, *57*, 131–32, *131*, 152
strong interaction, xiv–xv, 40
sulfur, 242–43, *243*
sun
energy from, xiv–xv, 181, 189, 190–96, 220
mass of, 195
nuclear fusion in, 193–95, 209–10, 220
supernova explosions, 34, 96–99, *98*, 252
Super Proton Synchrotron (SPS), *57*, *152*, 222, 224, 228
symmetry, between matter and anti-matter, 226, 235–38
synchronization, of clocks, 105–7

TASSO Collaboration, 240
telephone signals, 107–8
temperature, nuclear fusion and, 193–94, 208, 210–11
thermonuclear combustion, nuclear fusion and, 194–95
thermonuclear weapons, 210
thought experiments. *See* experiments, thought
time
absolute, 13, 18, 30–32, 73
as coordinate, 73–79, 159–60, 162
dilation of. *See* time dilation
as dimension, 73, 81, 163
direction and, 17–18
homogeneity of, 16–17
intuitive concept of, xiii
light and, 71
measurement of, 12–13, 18, 31, 100–102, 105–6
motion and, 15–17
muon decay and, *127*
nature of, 12, 136
Newton's concept of, 12–13, 30–32, 71, 89–90
relativity theory and, xiii, xv, 105–18, 159–60, 162
special theory of relativity and, xii
See also flow of time
time dilation, 107–18
equation for, 113–15
measurement of, 113–15, 121–33, 134–35
motion and, 108–13
observer and, 169–70
particle decay and, *156*, *157*
particle lifetime and, 155–58
relativity theory and, 107–18
research on, 120–33
space exploration and, 136–39
speed of light and, 111–12
twin paradox and, 139–43
time reversal invariance, law of, 231
timetable, space-time, 74
Treatise on Electricity (Maxwell), 101
Trinity College, Cambridge, 1–3, *2*, *5*
Trinity site (New Mexico), 202
tritium, nuclear fusion and, 210–11
tritons, nuclear fusion and, 210
Truman, Harry, 203
tunneling, *41*
twin paradox, 139–43

United States, 209
United States Naval Observatory, 135
University of Zurich, 55
upsilon particle, 239–40, *240*
uranium
bomb, 208
heavy, 198
in natural nuclear reactor, 214–15
nuclear fission and, 197–200, *198*, 212
Uranus, 7–8

vacuum, light traveling in, 91–92
variational calculus, 5–6
velocity. *See* speed
Voltaire, 7
volume, mass and, 10

wavelengths
energy of photons and, 46–47
light, 43
photoelectric effect and, 47, 48
radio, 43
X-ray, 43

wave theory
  light and, 38–39, 43–45, 64–66
  Principia and, 6
weak nuclear force, xiv
weak nuclear interaction, 122
Wheeler, John, 163
Wilson, Robert, *51*, 52
Woolsthorpe, 3–4

world lines, 76
  of light signals, 79–81, *81*
  particle lifetime and, 156–58, *157*
  space-time and, 76–79, *78*, 82
  twin paradox and, 141–42, *142*
World War II, 202–3

X-rays, 43